"This deeply insightful book highlights the profou[n]... ment is playing in catalyzing and shaping a new b..., ...it that will transform our economy and society. We are transitioning from passive consumers to active makers, driven by a desire to learn and achieve greater impact, and in the process rediscovering our humanity. If you want to understand where we are headed as a global society and why this is such a promising direction, this compelling and exciting book is a must-read." —**JOHN HAGEL,** founder and cochairman, Deloitte Center for the Edge

"*Free to Make* captures what it means to be human: to imagine, question, create, reflect, and try again. It's about making your own experiences matter and sharing them in ways that help make the world a changed place over time." —**MIKE PETRICH AND KAREN WILKINSON,** authors of *The Art of Tinkering*

"*Free to Make* is a comprehensive treatise on everything Maker. A leader of the Maker Movement since its inception, Dale Dougherty describes the roots of the movement and gives great examples of how it is changing lives and changing society. *Free to Make* answers the very important question: In today's society, where we can buy anything, why make? A must-read for any maker or anyone interested in becoming one." —**BRIAN KRZANICH,** CEO of Intel

This is a truly inspiring book by one of the great progenitors of the Maker Movement both here in USA and the world at large. Said most simply, we think with our hands as well as our heads—something we have forgotten in most of our current schooling. *Free to Make* provides a way to reach the many of us that find learning by sitting in a school room so boring. A sense of agency is the key to learning, and making things is a route to agency. —**JOHN SEELY BROWN,** former chief scientist, Xerox Corp and former director of Xerox PARC; coauthor of *A New Culture of Learning* and *The Power of Pull*

"*Free to Make* is a profound and joyful journey through a movement that is at once historical and profoundly contemporary. Imbued with sixties' sensibilities that give rise to creative acts of genius, whimsy, and passion, this book explores the ways in which the Maker Movement nurtures that irrepressible human desire to create and inspire others." —**MARGARET HONEY,** president and CEO of New York Hall of Science

Free to Make

HOW THE MAKER MOVEMENT IS CHANGING OUR SCHOOLS, OUR JOBS, AND OUR MINDS

DALE DOUGHERTY

with **ARIANE CONRAD**

Foreword by Tim O'Reilly

North Atlantic Books
Berkeley, California

Published by
North Atlantic Books
Berkeley, California

Cover art © iStockphoto.com/iconeer, iStockphoto.com/sweetjinkz,
  justone/Shutterstock.com; Makey Robot © Maker Media, designed by Kim Dow.
Cover and book design by Jasmine Hromjak
Printed in the United States of America

Free to Make: How the Maker Movement Is Changing Our Schools, Our Jobs, and Our Minds is sponsored and published by the Society for the Study of Native Arts and Sciences (dba North Atlantic Books), an educational nonprofit based in Berkeley, California, that collaborates with partners to develop cross-cultural perspectives, nurture holistic views of art, science, the humanities, and healing, and seed personal and global transformation by publishing work on the relationship of body, spirit, and nature.

North Atlantic Books' publications are available through most bookstores. For further information, visit our website at www.northatlanticbooks.com or call 800-733-3000.

Library of Congress Cataloging-in-Publication Data
Names: Dougherty, Dale, author.
Title: Free to make: how the maker movement is changing our schools, our jobs, and our minds / Dale Dougherty with Ariane Conrad; foreword by Tim O'Reilly.
Description: Berkeley, California: North Atlantic Books, 2016.
Identifiers: LCCN 2016011769 | ISBN 9781623170745 (paperback)
Subjects: LCSH: Makerspaces—Social aspects. | Industrial arts. | New products. | Inventions. | Creative ability. | BISAC: SELF-HELP / Creativity. | EDUCATION / Non-Formal Education. | TECHNOLOGY & ENGINEERING / Inventions.
Classification: LCC TS171.57.D68 2016 | DDC 600—dc23
LC record available at https://lccn.loc.gov/2016011769

1 2 3 4 5 6 7 8 9 UNITED 21 20 19 18 17 16

Printed on recycled paper

This book is dedicated to all makers, past, present, and future. My close friend Larry Martens passed away while I was finishing up this book. Born with deformed hands, Larry was a hands-on maker who loved to work with wood but enjoyed even more sharing his stories with friends and family.

Acknowledgments

I am fortunate to have the opportunity to learn from and work with so many clever, creative, and thoughtful people. I appreciate all the makers who have shared their projects and their stories with me and the world. I hope this book captures their enthusiasm, conviction, and optimism.

At O'Reilly Media, I've had such a long and productive relationship working with Tim O'Reilly, Laura Baldwin, Sara Winge, Gina Blaber, and Allen Noren. Thanks also to Christina Isobel, Arwen O'Reilly Griffiths, and Meara O'Reilly. I'd like to thank my current and past colleagues at Maker Media, especially Todd Sotkiewicz.

The Make Magazine Founding Team helped bring a rough idea into reality: Mark Frauenfelder, Phil Torrone, David Albertson, Shawn Connolly, Dan Woods, Keith Hammond, Goli Mohammedi, Michael Shapiro, and Kirk Rohr. Also, Paul Spinrad, Daniel Carter, Jason Babler, and Gareth Branwyn have played important roles on the magazine, and now Mike Senese is at the helm.

For over ten years, the core Maker Faire Team, led by Sherry Huss and Louise Glasgow, has created and shaped a new kind of participatory event. Along with Bridgette Vanderlaan, Jonathan Magnin, and Kate Rowe, they bring passion, dedication, and caring to producing an event that is challenging to do. There are many other team members who contribute behind the scenes, making these events amazing and safe for everybody. I also want to thank Sabrina Merlo and the producers of global Maker Faires, who bring the same kind of dedication to their local communities.

I'd like to thank the team at Maker Ed for their efforts to share with educators how to engage more children in the practice of making. That includes Warren "Trey" Lathe, the executive director, and his team as well as the board members and funders. I'd also like to thank AnnMarie Thomas for helping get the nonprofit up and running.

My collaborating author, Ariane Conrad, helped me organize my thoughts and my writing, while contributing her own insights on the Maker Movement.

I am grateful for the love and long-standing support of my wife, Nancy, and my entire family—especially my children, Katie, Ben, and Glenda. Thanks also to my mom.

To all of you who have shared all or part of this journey and hold on to its promise, I hope we continue to share the belief that all of us are makers and that the future will be better only if we participate in its creation.

Contents

Foreword

Dale Dougherty makes some of the best beer I've ever tasted. Each year, he makes events that are attended by hundreds of thousands of people. He makes a magazine. But above all, he makes a map.

Before Dale launched *Make:* magazine in 2005 and Maker Faire in 2006, the word *maker* existed, but it was not a badge of pride or identity. After *Make:* appeared, the word was everywhere, and people as diverse as electronics hobbyists, crafters, mechanics, artists, and hardcore engineers proudly began to refer to themselves as Makers.

I still remember scenes from the first Maker Faire in 2006, the collision of communities that were now suddenly united by a name, a mirror that, with wonder, they saw themselves in. There was the armorer from the Society for Creative Anachronism, who makes chain mail for medieval pageants, deep in conversation with the teacher from the Crucible, where inner city kids learn to weld. There was the Swap-a-rama-rama, a clothing swap augmented with sewing machines and silk screening machines, where people remanufacture clothes and have a fashion show at the end of the day, right next to the Alameda Contra Costa Computer Recycling Society, showing off their biodiesel-powered supercomputer running Linux on a cluster of recycled PCs. Segway Polo, fire-breathing cars and metal sculptures, human-powered amusement-park rides, craft beers and 3-D printers, robots and music—all tied together by the proud assertion that "I made this."

The poet Wallace Stevens wrote "said words of the world are the life of the world." Language is a map that helps us see, and the people like Dale who tell stories about what is true and how to see hidden connections are its mapmakers.

Make: magazine was not Dale's first exercise in mapmaking. Working with me, he helped to write and publish what are still widely considered the best books on computer programming, books that quite literally justified the line on the cover of *Publisher's Weekly* in 2000,

"The Internet Was Built with O'Reilly Books." In 1993 he launched the world's first commercial website, which he quite aptly named the Global Network Navigator, or GNN, in homage to *The Navigator,* an early-nineteenth-century guidebook for those descending the Ohio River from the rivers of the eastern seaboard. GNN was a catalog of the world's best websites and an online magazine about the people who built them. Dale also organized the World Wide Web Wizards' workshop in 1993. Held at the O'Reilly offices, this meeting brought together the leaders of the nascent web, Tim Berners-Lee, Marc Andreesen, Rob McCool, and others who went on to build world-changing technology.

Ten years later, Dale coined the term "Web 2.0" to describe the second coming of the Web after the dot-com bust, and once again captured the zeitgeist of a generation of technologists. Why did some companies survive the bust and thrive? What were the rules of success in this new medium? Together we created a story that helped to revive the industry and showed the path forward.

So when Dale leaned over to me in a taxicab one day in 2004, and pitched me on his latest idea, a magazine about the people who, he believed, were going to shape the next big wave of technology, it took me only a moment to say "yes!" The same kind of people who ten years earlier had been exploring the Internet were now exploring the new world of cheap hardware, sensors, and new kinds of manufacturing. They were taking devices apart and rebuilding them. They were using the vast graveyard of expired consumer electronics as a parts shop for invention. They were imagining and building new kinds of tools. But most of all, they were learning and having fun.

Make's slogans—"Technology on your own time" and "If you can't open it, you don't own it"—became the manifesto of a new movement. Every quarterly issue of *Make:* was eagerly awaited by everyone from harried parents to high-tech executives (an overlapping set). Maker Faire, launched the next year, brought them together in a celebration of their shared passion. But more than that, it became "the world's biggest show-and-tell," as Dale likes to describe it today, and the heart of an education revolution that is inspiring a new generation to care about science and technology not as "the thoughts of a dry brain in a dry season" (per T. S. Eliot) but as the living, breathing exploration of endless possibilities.

When twelve-year-old inventors, not just athletes, musicians, and movie stars, are hosted by the president at the White House, that signals to a whole generation that their innate love of exploration and play need not be put away, but should instead be the foundation of their future aspirations.

Dale's infectious enthusiasm, his ability to build communities by showing the hidden connections between technologies and bringing their inventors together in new combinations, but above all, his passion to teach and inspire, shine through in this book.

Find your own inner Maker. And start to measure yourself by what you make, not what you own or buy.

—*Tim O'Reilly, founder and CEO, O'Reilly Media*

Introduction

Nobody needs to make anything today. Necessity is no longer the mother of all invention. We can find almost everything we need and buy it, more easily and more cheaply than ever before. And just as easily, we throw away those things.

Yet today many people are making what they need, and there are more of them than I ever imagined when I first called them out as "makers." What—or rather who—is a maker? Makers are producers and creators, builders and shapers of the world around us. Makers are people who regard technology as an invitation to explore and experiment, with the most inclusive possible definition of *technology,* meaning any skill or technique that we learn and employ. What we once called hobbyists, tinkerers, artists, inventors, engineers, crafters—all of them are makers. The power of "maker" as a new term lies in its broad application, its sense of inclusiveness, and its lack of close alignment with a particular field or interest area, so people are free to claim the identity for themselves.

Makers share a mindset, and that's why the Maker Movement has emerged as a global countercultural phenomenon, inviting everyone to join in and make something. It is this generation's rock-and-roll, a contagious DIY spirit that echoes the "do your own thing" mood of the 1960s, captured in the song by The Mamas and The Papas that says, "Make your own kind of music / Sing your own special song." The Maker Movement is a platform for creative expression that goes beyond traditional art forms and business models. It is a collaborative form of problem-solving, from the practical to the hypothetical, leading to new products, new ways of learning, and new ways of doing science. It is an intergenerational movement, bringing together young and old, as it has for many generations before us, to discover and learn about the world.

By analogy, almost nobody needs to cook—there are plenty of ways to buy prepared food and fast food. (I will use a lot of cooking analogies in this book, because cooking is my own favorite kind of making.) Yet those of us who find cooking enjoyable develop the skills and knowledge to become even better everyday cooks. Cooking a meal with family and friends can be about as satisfying an experience as life offers. What's more, cooking for yourself gives you control over what you eat, and generally, it is much healthier because you know what you are eating. Cooking is something you are free to do. There are lots of ways to learn how to do it.

Making can be hard to do. It requires skills, knowledge, and tools that some think are beyond them. That represents a barrier for many people, but it also provides a challenge that is more rewarding when we succeed. Making today is easier than it ever was. There's better information available, and a community that shares its expertise openly online and in more and more accessible makerspaces. The tools are better, with software reducing the complexity of many tasks. If you want to make, there has never been a better time.

Why make? For one thing, making is something humans are wired to do. It comes naturally to us, using our hands and minds to create, explore, and communicate. It is something we can do once we have an idea about what we want to do. Making is a kind of "what if," exploring questions as to how something works and if it could work differently.

Making leads to innovation. We may see the need to personalize something that we can't buy, or fix something we can't replace, or discover an opportunity to create something new that does not yet exist. We can make it. Innovation doesn't have to mean that you've created the next big thing. We can innovate usefully in many small ways that are also important. Making is also a process that combines play and learning. When people make things, they enjoy the process of making, both its frustrations and its rewards. They can take pride in the result—something they can hold in their hands and show to other people. They also take pride in what they've learned. Deciding to innovate or choosing to make for its own sake are both ways we can make change, changing our own lives and impacting the lives of others.

Making is a creative freedom that we learn to appreciate through practice. We can get better at it, and as a result, there's even more we can do. We don't know if we have any special talent for it unless we get our hands dirty and try. Through the practice of making, we develop what we once called a "can-do" mindset that encourages us to act, take control of our lives, and develop our own capabilities. Making engages us fully and deeply as human beings, and it satisfies our creative souls. Maybe making can change the world, but first it changes us. We begin to see ourselves as confident, capable, and creative individuals.

Today, all of us are consumers. In consumer culture, we define ourselves based on what we buy or own. Consumer culture disables us in some ways, and people can feel entitled, expecting others to do things for them. (Reading reviews on Yelp makes me think that nobody is ever pleased.) This kind of consumerism disconnects our desires from our own work, the work that is required to realize our desires. We are often left unsatisfied and unfulfilled, perhaps not even knowing what we truly desire. In the extreme, consumerism is a form of learned helplessness. In consumer culture, making is something that we've forgotten we can do. It has pushed making from the mainstream to the margins. However, there's something else available to us: We can see ourselves as producers. In maker culture, we define ourselves based on what we can do, and what we can learn to do.

It is the goal of this book to convince you that all of us are makers. The joy of making is greater than anything you can buy. Makers are playful, resourceful, and experimental. They not only help themselves but they help others. Makers do well because they can adapt to change as self-directed learners, but they are also agents of change. In this movement, makers are setting their own agendas and asking others to join them.

In this book, I will be your personal tour guide to the people and projects that are the evidence of the Movement, in much the same way I escort guests and journalists through Maker Faire. We dubbed the Maker Faire, a global event I founded in 2006, "The Greatest Show-and-Tell on Earth," which captures something of the circus-like atmosphere that annually draws hundreds of thousands of adults and children who come to marvel at the endless imagination and creative capacity of inventors and crafters, engineers and artists of all ages, who display projects that blur the boundaries of technology and science, craft and art. There are homemade autonomous robots, banana pianos, motorcycle jackets glowing with LEDs, motorized lounge chairs, ale brewed from fossils, a fire-breathing scrap-metal octopus, kite-mounted aerial photography rigs, self-watering gardens, wind turbines made of bicycle wheels, and thousands of other playful and practical projects.

Makers are people making all kinds of things for a variety of personal, social, and commercial reasons and for educational, artistic, practical, commercial, or entertainment value. I'll introduce you to some of these makers and let them tell you their own stories. You'll meet Lisa Qiu Fetterman, who created a new, more affordable sous-vide cooker; Brook Drumm, a former minister who created his own 3-D printer company; Quin Etnyre, who started his own maker-kit business as a twelve-year-old; Marque Cornblatt, a performance artist who started a drone racing league; Pam Moran, a school superintendent who is organizing all the schools in her district around maker culture; Ayah Bdeir, who turned LittleBits from an academic project into a successful line of products that introduce children to electronics; and Nick Pinkston, who is creating the fully automated factory of the future.

Some makers are professionals, but many are not. They are amateurs doing something they love to do. That passion is what first attracted me to makers. Sure, they are smart, clever, and curious, but what I find appealing about makers is that they are more like kids than adults, playing and having fun. I am privileged to have created a context in which I constantly get to meet makers when I launched *Make:* magazine in 2005. Over the past decade I've witnessed a worldwide renaissance of creating, designing, modifying, inventing, customizing, and personalizing.

I sometimes wonder why I "discovered" makers and became so fascinated by them. I am not an engineer or designer. I wasn't the kind of person who from birth understood himself to be a maker. I was an English major in college and I saw myself as a creative person. What I knew about myself was that I loved to learn. I learned not just from books but from people, especially those who had superpowers I wished I had.

Informally, I acquired something of a technical education. I don't mean that I gained extensive technical knowledge, but that I learned to solve technical problems and enjoy the process. More importantly, I became comfortable talking to people who were a lot more technical and a lot smarter than me. By following talented people and being a part of technical conferences and workshops, I was able to see how technology was changing our world and to understand that the opportunity to be part of that change was open to just about anyone, including me.

The Internet, the World Wide Web, and open source all started out as technical communities outside the mainstream. The company that Tim O'Reilly founded and where I got my start was a publisher of very technical books. We also began to play an important role in convening emerging technical communities that were self-organizing rather than hierarchical. We learned to recognize and trust new patterns before anyone else did by following what people were interested in and the problems they were working on. We also understood that the collaborations among contributors were creating real value and eventually impacted all of society.

After the tech crash in the early 2000s, I had two pretty good insights. One was Web 2.0, which led to a conference by that name. The other was the idea that hackers were modifying the physical world, wanting to hack their cars the way they hacked their laptops—to make changes, to personalize and customize it, and to get the physical world to respond to them in new ways. It really was a thesis, but when I started talking to people about making, nearly everyone responded by telling me what kind of project they were working on. If they weren't making, they talked about a brother or aunt who was a maker. I was struck by how personal each response was, and how much enthusiasm was behind each story I heard. As I met more and more makers, I believed that they were a newly forming technical community that was outside the technology mainstream even of Silicon Valley and outside the mainstream of consumer culture.

Energized by the Internet and increasingly affordable technologies for design and production, the maker community has grown to include more than just its technical members. It has become a participatory movement for creating a future that, to paraphrase William Gibson, is here now but not yet widely distributed. Thus, the Maker Movement is not about the technology itself, but about people and their projects that apply technology in new ways. A. P. Usher, in *A History of Mechanical Inventions,* discusses how invention is about establishing relationships that did not previously exist. This book is about the relationships that we form around making, and the communities that form around tools, knowledge, and materials. It's about local maker communities in San Francisco and New York, Detroit and Dallas, Paris, Barcelona, and Shenzhen, China.

The Maker Movement is changing who gets to make things, what gets made, and where and how things are made. It is a prototyping revolution that seems to follow from the desktop publishing revolution, allowing more people to turn an idea into a tangible object. Economist Jeremy Rifkin called it "the third industrial revolution"; *Wired* editor Chris Anderson called it "the new industrial revolution." However, it's not a revolution that will see more people working in factories. Instead more of us will own or have access to the equipment that a factory might have, as one might have access to equipment at a gym. More than an economic change, the Maker Movement is a cultural shift that is leading to a creative flourishing of art and science, technology and craft, a hands-on renaissance that is producing new tools and new ways of thinking.

There are a number of questions we will explore. Why is making resonating so deeply and for so many? Why now? How does making impact our personal development and our ability to learn, to thrive, and to work with purpose and dignity? How is making changing the experience of learning for children and adults as well as transforming our school system? How are makers developing innovative new products? How are makers getting things made, manufactured, and distributed? How are makers serving the needs of others in the community? How could making empower us to solve the social and economic challenges we face as human beings around the world?

Making is at the confluence of the democratization of technology and the democratization of learning. Stated another way, the means of production, and the knowledge and skills required to make use of them, are available more broadly—to just about anyone. Technology is cheaper and faster, as is learning itself. Nothing is truly free—as in free beer—but it is there at our fingertips. Information is accessible, independent of institutions or corporations that in the past have had the ability to control access.

The Maker Movement signals a societal, cultural, and technological transformation that invites us to participate as producers, not just consumers. It is changing how we learn, work, and innovate. It is open and collaborative, creative and inventive, hands-on and playful. We don't have to conform to the present reality or accept the status quo—we can imagine a better future and realize that we are free to make it. Making is the mother of all possible futures.

"We're given a chance right now to launch a tremendous experiment in which we'll do things almost totally differently from the way they have been done before."[1]

That's how Historian Gary Wills expressed Thomas Jefferson's attitude toward revolutionary change.

Embracing that same revolutionary attitude is what the Maker Movement is all about.

1

We Are All Makers

Making is so central to what makes us human that the term *Homo faber* was coined to describe what sets us apart from other animals. Latin for "man the maker," *Homo faber* goes back to the writings of a Roman politician who lived around 300 BCE, Appius Claudius Caecus. He oversaw the building of the Appian Way, essentially the first long-distance highway, and the Aqua Appia, the first water supply system for Rome. The term *Homo faber* reminds us that humans both transform materials into tools and transform the world using those tools. This makes us who we are: makers.

Caecus's concept of *Homo faber* was one of many ideas rediscovered from Greek and Roman antiquity during the Italian Renaissance. At that point, "man the maker" acquired the further meaning of "man the creator." The act of creation was essentially a divine power. With the ability to do ourselves what the gods could do, we derived a sense of mastery of the world. Artists like Leonardo da Vinci and Michelangelo represented a new, creative spirit that transcended limits and transformed the world. With the invention of the printing press by Gutenberg, and a scientific revolution spurred by Galileo, the Renaissance was a golden age of art and innovation, a time when tradition was challenged and new ideas flourished.

While humans are not alone in using tools, no other animal makes them. Novelist Bruce Sterling, in the first issue of *Make:* magazine, wrote a column called "Make the Tools That Made You" on the subject of flint knapping—shaping stone tools. "Before modern humans came along, our Stone Age ancestors spent 1.8 million years making tools out of rocks," writes Sterling, adding, "It follows that you, a modern human, have evolved superb eyes and hands for that job ... [and in fact can] make better Stone Age tools than our Stone Age ancestors." Although making stone tools by hand is hard work, flint knapping continues today as a hobby.

Making is in our bones, in our blood, in our DNA. We are born makers, with an innate ability to make with our hands, which allowed our ancestors to adapt and survive. We learned to make from each other, even across generations, by using that other uniquely human skill: language. As Adam Savage, cohost of the popular television series *Mythbusters,* said at Bay Area Maker Faire: "What distinguishes us as humans is our ability to make tools and tell stories." Making and talking shaped our brains and our social lives over centuries.

In modern times, humans were classified as *Homo sapiens,* initially used to distinguish us from apes. We might think of *Homo sapiens* as "making knowledge," a creative or productive act in itself. We can also see our ability to make tools and to make knowledge as entirely compatible. The historian and philosopher Mircea Eliade wrote that the maker is the knower. It isn't just what you know, but what you could do with your knowledge. Writing about his theory of multiple intelligences, Howard Gardner notes that one form of intelligence is "the capacity to solve problems or to fashion products that are valued in one or more cultural settings."[1] There's an intelligence that expresses itself not as IQ or as test scores or rankings, but in the quality of problems solved, or the products that we produce. We see that kind of intelligence in the work of artists, inventors, and innovators.

In his book *Sapiens: A Brief History of Humankind,* Yuval Noah Harari claims that tool-making is not what makes *Homo sapiens* different from other human ancestors.[2] Instead, he describes a cognitive revolution—an information revolution thanks to language—that took place just over one hundred thousand years ago. Language evolved as a way of sharing information about each other, and much of it took the form of gossip to keep tabs on what others were doing. Sharing information helped us develop techniques for cooperation—working together to solve problems. Harari emphasizes that language was used not just to describe reality, but also to talk about a world that could be imagined, by telling stories and painting pictures.

If *sapiens* is reasoning and *faber* is making, writes Johan Huizinga in his book *Homo Ludens*, then there is a "third function … just as important as thinking and making—namely, playing."[3] I read Huizinga after college and was struck by his focus on something that seemed so core to human experience. I hadn't encountered anything like what Huizinga expressed about play, and I haven't since. (Perhaps it did not seem a subject for serious study to others.) Huizinga writes that "the fun of playing resists all analysis, all logical interpretation,"[4] which is why he set out to study "the play-element in culture."

Play is special, somehow outside ordinary life, beyond the necessity of work or meeting our basic needs. It is a voluntary activity. We choose to play, and we really cannot be forced to do it. "Child and animal play because they enjoy playing, and therein precisely lies their freedom.… The first main characteristic of play is that it is free: is in fact freedom."[5] And yet, he writes, it "means something." Play can be seen as a celebration of what we value. The celebration itself binds people together, and is therefore central to the formation of our culture. "Civilization arises and unfolds in and as play,"[6] writes Huizinga.

Humans are tool-makers, inventors and innovators, storytellers, tinkerers, and role-players. We are makers who are free to imagine, free to play, and free to make.

AMERICAN MAKER

In 1961 a short industrial film, created by Chevrolet, opened in American theaters. Its opening shot is of the Pacific Ocean. A close-up scans a large rock outcropping to reveal behind it the beach where two boys are making a sand castle in the distance. The title appears to great fanfare: *American Maker*. The narration begins, following close-ups of the boys building the elaborate sand castle:

Of all things Americans are, we are makers. With our strengths of minds and spirit, we gather, we form, and we fashion. Makers, shapers, put-it-together-ers. We start young, finding out early in life what it's like for something to grow and take shape beneath our hands. We start young and stay young, modeling with careful pride the things we expect to endure for ourselves and for others. We build for use and we build in fun, joining eyes and hands and brains.[7]

It was still commonplace for Americans to think of themselves as makers. Making was a source of pride, personally and for the country—and not just in the United States. After a talk I gave in Rome, a man came up to me and excitedly said, "We don't have the word *maker* in Italian, but we are makers." He wanted me to know that making defined Italians as a people and as a country. It spoke to their Roman past of roads and aqueducts, as well as their commercial reputation in design represented in cars, furniture, and clothing.

I used the opening of *American Maker* in a TED Talk I gave at the Fox Theatre in Detroit in 2009. As I prepared that talk, I realized that I could easily say that some people are makers. I knew from people I met through the magazine and Maker Faire that some people clearly identified as makers from the time they were born, even if that wasn't the name they called themselves. Yet, I wondered if I could say that not just some of us were makers, but all of us were makers. To be honest, I didn't know if it was true, but I believed it to be true by this logic: if making was core to what makes us human, then all of us makers. I acted on my belief, and I felt the talk I gave in Detroit, seen online by many others since, was inspired. This idea, this belief, still resonates with us, that seeing ourselves as makers is what it means to be human.

Popular Science magazine started in 1872 and faced a number of struggles finding an audience for science journalism, but it eventually lived up to its name and became a popular magazine read by millions. The magazine said it was for "the home craftsman and hobbyist who wanted to know something about the world of science." One issue said it covered inventions, mechanics, money-making ideas, and home-workshop plans and tips. The covers of the magazine often featured futuristic flying machines—the flying car has been its most popular image, and was repeated regularly. What became *Popular Science* wasn't about science; it was about you. It was about what you could do or would be able to do. The future was waiting for you to figure out how to get there.

I have found old copies of *Popular Science* and *Popular Mechanics* at a flea market. A copy of *Popular Mechanics* with "What to Make" on the cover, dated 1944, cost fifty cents. When I flipped through these often moldy copies, some of which dated back to the early twentieth century, I was struck by the practical nature of the stories, and yet how they covered a wide range of projects, some of which might be considered whimsical or eccentric. A story on the construction of a two-car garage out of masonry might begin, "Building your own freestanding garage might seem like a challenge too big and taking too long to be practical. Yet I did it, and I'd like to tell you how you can do it too." Several pages later would be a story about building a fancy three-story birdhouse for martins, and then a wind-powered ice-sailing boat for frozen Midwestern lakes. These stories were accompanied by terrific black-and-white illustrations, which were often clearer than the photographs. More than anything, I loved that anyone who was willing could do these projects. The majority of articles were submitted by readers and conveyed their own pride in sharing what they had accomplished. I was impressed by the "can-do" attitude. I realized that many of the projects in those magazines would be fun to do today.

In these magazines, making things was ordinary and normal, not unusual. There were plenty of good reasons to do projects, including saving money and enjoying a hobby or improving your home. These articles were not only read by adults; they inspired young people to experiment and do things.

Robert Noyce, a cofounder of Intel in Silicon Valley, was born in 1927 in Grinnell, Iowa. The author Tom Wolfe interviewed him and wrote an article in 1983 in *Esquire* titled "The Tinkerings of Robert Noyce: How the Sun Rose on the Silicon Valley." Wolfe wanted to understand why so many of the giants of the Information Age came from small cities in the Midwest rather than from large cities. Wolfe describes Grinnell as the essence of the Midwest: "It was one of the last towns in America that people back East would have figured to become the starting point of a bolt into the future that would create the very substructure, the electronic grid, of life in the year 2000 and beyond."

Noyce was one of four sons of a minister. He was a good student who showed an aptitude for athletics and science. He did odd jobs and was a Boy Scout. Yet the story that Wolfe tells is of a thirteen-year-old Noyce and his brothers building a box kite that supposedly could lift a human off the ground. They got the idea and the plans from the *Popular Science*. Bob Noyce and his brother Gaylord built the kite and then tried to fly it. Running through a field was not enough to get the kite to lift off the ground, so they took it up on the roof of the barn. "Bob sat in the seat and Gaylord ran across the roof, pulling the kite. Bob was lucky he didn't break his neck when he and the thing hit the ground. So then they tied it to the rear bumper of a neighbor's car. With the neighbor at the wheel, Bob rode the kite and managed to get about twelve feet off the ground."[8]

This is what you did in small towns in America. You learned to see what you could do with what you had. Wolfe asked Noyce directly why he thought that small towns produced so many good engineers. "In a small town," Noyce liked to say, "when something breaks down, you don't wait around for a new part, because it's not coming. You make it yourself." People had no choice but to fix things and figure out creative ways to meet their needs. They were resourceful. Whether you worked on a farm or a factory, lived in a city or a suburb, the ability to tinker and make things better was something you practiced.

Many people still fondly remember the *Heathkit* catalog. The Heath Company, based in Benton Harbor, Michigan, sold kits to make electronic products like radios and amplifiers. According to an excerpt from the original *Heathkit* catalog, "anyone, regardless of technical knowledge or skills, could assemble a kit himself, and save up to fifty percent over comparable factory-built models. All that would be required were a few simple hand tools and some spare time."

For the article "The Soul of an Old Heathkit,"[9] I interviewed Howard Nurse, who not only grew up on Heathkits; his father became president of the company. He told me that electronics were not readily accessible in 1950s. The only place he could see electronic components was at a local TV repair shop, which he hung around. The *Heathkit* catalog opened a door to the new world of hi-fi components, electrical test equipment, ham radios, and later, television sets. He recalls the joy of opening up the box. "First, you'd see the Heathkit manual, which was the heart of the kit." Then you'd find the capacitors and resistors in brown envelopes. A transformer came wrapped in spongy paper, a predecessor of bubble wrap. "Before you did anything, you had to go through the errata that came with the kit." Then he would do an inventory of the parts, using a muffin tin to sort them. Additionally, he'd use corrugated cardboard to arrange the small capacitors and resistors in rows. "After all this waiting and preparation, you'd begin to assemble the parts," he said. "You started by attaching a few components, and then you got to solder, which was really fun … flux[10] was an aphrodisiac."

When you finished the assembly of a Heathkit and tried it, often it didn't work. This too was part of the process of understanding electronics and learning to fix problems. Nurse built his first computer from a Heathkit, the H8 digital computer. It's ironic that the Heathkit computer, released in the 1970s, was the culmination of DIY electronics, but the rise of ready-made computers killed it off—and eventually killed the Heath Company too.

The era from the late 1940s through the 1970s, when *Popular Science, Popular Mechanics,* and Heathkits were in their heyday, might be considered a golden age of tinkering in America. It was a middle-class virtue that enabled the people to have things and do things that they needed or wanted but couldn't afford, like expanding the house for a new family member, sewing clothes, or repairing the car because they couldn't afford a new one. What you could do for yourself and your family was a source of pride.

ARE WE MAKERS ANYMORE?

Do Americans still see ourselves as makers? Consider the 2008 short film *Brighter*. It opens with a shot of the elevators inside an upscale department store against the soundtrack of plaintive guitar strumming. "We are a nation of consumers," begins the soothing narration, "and there's nothing wrong with that. After all, there's a lot of cool stuff out there …" The film was brought to us by Discover Card.

In the course of a few decades, America evolved from a nation of makers into a nation of consumers. Americans worked more hours and had fewer hours available to tinker. We chose to pay professionals to transform the garage into an in-law suite rather than do it ourselves. We became detached from the questions of how our food, cars, electronics, toys, shampoo, and so on were made, where they came from, and who produced them. The production and distribution of goods and products became less and less visible to us as factories were outsourced to places like Mexico, Bangladesh, and China, eliminating American working-class jobs. Advertising, which really took off with television in the 1950s and 1960s, persuaded us that the more stuff we bought, the happier we'd be. Ads could create demand for more and more products, which we might call manufacturing desire.

Industrial production and consumerism meant trade-offs that were usually about getting the smallest number of people and the lowest costs involved in the process to give those industries the maximum benefit. Industrial producers made—have been making—a bunch of choices that we have had no ability to influence and with which we might not agree. At the same time, a message has been broadcast to us, every hour of every day, about our personal freedom equating to the wide range of choices we have in the aisles of stores. Becoming a "smart consumer" meant being able to navigate all these choices and make the best decisions. Benjamin Barber, a professor and political theorist, wrote in his book *Consumed:* "We are seduced into thinking that the right to choose from a menu is the essence of liberty, but with respect to relevant outcomes, the real power, and hence the real freedom, is in the determination of what's on the menu."[11]

There are more things available to us than ever. The selection is endless on Amazon or in Walmart. In *Sapiens,* Harari cites the scarcity of artifacts in the life of hunter-gatherers compared to their modern descendants. "A typical member of a modern affluent society will own several million artifacts—from cars to houses to disposable nappies and milk cartons." He adds: "There's hardly an activity, a belief, or even an emotion that is not mediated by objects of our own devising."[12]

Yet, in fact, "our own devising"—making—and our consuming are mostly walled off from each other. It is the difference between eating a meal that you've cooked and eating fast food, having little involvement or knowledge about the ingredients, processes, or people involved in the preparation of the food. We overeat and are still not satisfied. Just being consumers does not seem to satisfy us. We might feel empty, disconnected, even helpless. Mostly we are bored.

Can creative acts such as making things make us feel more alive? And if we surround ourselves with objects that have a true connection to us and to our efforts, might we feel more fully alive?

UNDERGROUND MAKING:
PUNKS, HACKERS, AND HOMEBREW COMPUTERS

As consumerism became dominant, making went underground and over the next fifty years emerged mostly in subcultures. One example was punk rock, which subverted the corporate production of music by bringing music back to its roots—the garage band. It didn't really matter how good you were musically. What was important was that you made your own music your own way. Punk bands recorded, produced, distributed, and performed their music themselves without mainstream music labels or corporate sponsors. They were really the earliest adopters of the do-it-yourself, or DIY, ethic.

Fashion could be punk too: in practice safety pins, Doc Martens boots, and black everything, but philosophically individualistic and transgressive, often defying gender norms. Legendary musician Patti Smith, sometimes called the godmother of punk, said, "To me, punk rock is the freedom to create, freedom to be successful, freedom to not be successful, freedom to be who you are. It's freedom."[13]

This subversive notion of DIY could describe people deciding to become self-sustaining by growing their own food, to learn how to repair their own Volkswagens, and to sew and repair their clothes rather than buying new ones. They were nonconformists exploring alternatives, seeking freedom by not accepting what was easily available.

Another subculture that was finding its own way was grounded in hacking. The book *Hackers* by Steven Levy opens with the story of Peter Samson and the Tech Model Railroad Club at MIT in the early 1960s. Levy writes that Samson and his friends "had grown up with a specific relationship to the world, wherein things had meaning only if you found out how they worked. And how would you go about that if not by getting your hands on them?"[14] They were tinkerers.

Levy identifies this as the "Hands-On Imperative," one of the tenets of the Hacker Ethic. The Tech Model Railroad Club consisted of students who, it might be said, never grew up. They continued to be fascinated by toy trains; one group worked on the landscape and train layout, while Samson's group specialized in the switches that controlled the model trains. Out of this latter group emerged the first hackers who were fascinated by how computers work. They recognized the potential of computers as tools for their own purposes. They wanted to get their hands on the computers; they needed the time to explore; they didn't want to go through a centralized bureaucracy to have computing services performed for them. "Hands-on" was synonymous with access to learning directly how to do things yourself. They began forming a set of ideas that computers should be open and accessible systems, and the information about how they worked should be shared freely.

From its earliest beginnings, hacking was about the freedom to personalize technology. It didn't matter whether the technology was designed to do what they wanted it to do. What mattered was whether the system was open and flexible enough to do what they wanted. It's not a surprise that the free software movement comes out of MIT, derived in part from the people and experiences of the Tech Model Railroad Club.

There was another group on the West Coast playing with technology, specifically trying to design and build computers themselves. This was the Homebrew Computer Club in Menlo Park, California, which started in 1975. Lee Felsenstein, who created the Osborne Computer, opened meetings by saying "Welcome to Homebrew Computer Club, which doesn't exist." Their offbeat humor was one of the things that united them. Unlike computer "user groups" that would come later, these hobbyists got together to swap parts and share information about building their own computers. Club members Steve Wozniak and Steve Jobs met in high school and, according to Wozniak, they had "two things in common: electronics and pranks."[15]

Woz said of his time there: "I just loved going down to the Homebrew Computer Club, showing off my ideas and designing neat computers. I was willing to do that for free for the rest of my life."[16] That enthusiasm and motivation, being driven only by the desire to "scratch their own itch," as Eric S. Raymond described it in his book *The Cathedral and The Bazaar*, became another hallmark of the hacker community.[17]

FROM MARGINAL TO MAINSTREAM: PERSONAL COMPUTERS

On a weekend in April 1977, the first West Coast Computer Faire opened to unexpected success. Its founder, Jim Warren, called it "a mob scene" with over twelve thousand people attending. The goal of the Faire was to bring together hobbyists who were making homebrew computers. Warren said that the West Coast Computer Faire was like one of the "happenings" in San Francisco in the 1960s: "Back then it was power to the people; now it's computing power to the people."[18] Steve Wozniak and Steve Jobs were there, showing off the computer they built in a garage. The Apple Computer exhibit displayed the new Apple II. Mike Markkula, then vice president of marketing at Apple, said of the Faire: "I'm not exactly sure why so many people are here. A lot of them are just curious about what's going on."[19]

The West Coast Computer Faire showed that what hackers were doing was not limited to just the members of the Homebrew Computer Club. It was increasingly of interest to more and more people who did not consider themselves hackers. Wozniak estimates that twenty-one companies could trace their roots back to the Homebrew Computer Club.

There's one more story to add, of two young geeks named Bill Gates and Paul Allen. After reading an article in *Popular Electronics* in 1975 about the Altair computer, Allen persuaded Gates to drop out of Harvard. They went to New Mexico to meet Ed Roberts and began writing BASIC, the first programming language for a personal computer, and later developed an operating system, DOS. Despite having no formal training, they spent unlimited time learning what this technology was all about, and ultimately understood that they were creating a new world of personal computing.

Today, nearly everyone has a computer, but back in the 1980s, as the first personal computers emerged, not many people had a good understanding of why they would want one. Many people thought that computers were fancy calculators or glorified typewriters. Computers for data processing were needed in corporations but not at home. The personal computer needed a killer application, and it turned out that it was not about computers as number crunchers, although they could certainly do that.

In 1985, Apple came out with the first LaserWriter, a laser printer. Coupled with page design software, Macintosh computers could be used to design a document and then print it. WYSIWYG (what you see is what you get) page layout programs were the first of a series of computing applications with broad appeal that allowed people to see the computer as a creative tool. It enabled them to do something that was either hard to do manually or required special expertise. The combination of software and hardware created a desktop publishing revolution, where anyone could design and print typeset-quality brochures, newsletters, and books. Designers and architects began to use computers to do what they did manually on a drafting table. Computer-aided design (CAD) was born. Those who learned to design for laser printers soon began designing for digital media, first for video games and multimedia CD-ROMs and then the Web. A new set of creative industries was born.

Desktop publishing is an early example of the democratization of technology, as the typesetting and printing business was disrupted and amateurs could now do what only trained professionals formerly could. It didn't necessarily mean that what they produced was good, but many learned to use the tools and get better at them, motivated by the amount of control they had.

Riding the punk aesthetic into the world of the printed word and literature, boosted by the availability of desktop publishing, came the birth of zines: small-batch, homemade, low-budget underground magazines created by and for geeks with unusual interests outside the status quo. The design of zines was often purposely sloppy to emphasize their oppositional stance to mainstream media. Vibrant communities often formed around them. When the Internet became mainstream, a lot of the energy of zines and their communities transferred online, to bulletin board systems and then blogs.

All of these developments rested on the newfound ability to create and distribute content without any oversight or middleman—the printing press 2.0. A personal computer could now act as a recording studio, a publishing house, a TV production unit, or the doorway to a globally active bulletin board system. Making—more and more in digital forms such as music, publications, and interactive content—was about engaging communities of enthusiasts, freedom of expression, and personal choices.

HACKING AS A LIFE SKILL

Today, the word *hacker* has both positive and negative meaning. The media often paints the hacker as a miscreant, someone who breaks into computers and steals data. But along the way, the term also made a leap into the broader cultural meme pool. Sometime in the 1990s, people began talking about "hacking" outside of computing: there were food hackers and financial hackers; people were sharing "hacks" on how to book airline travel, how to become more productive, or how to parent. In the self-service economy of the Web, life hacking was becoming a valued skill. Hacking was how you got what you wanted.

At O'Reilly Media, I paid attention to what hackers were doing from the time I began to write Unix manuals in the 1980s. I didn't see hackers as a personality type or a lifestyle. I saw them as people who came up with clever or non-obvious solutions to interesting problems. I realized that what hackers were doing was important on a number of levels. They represented a fundamental shift in the way we think about how things are made, and how people work together.

Starting in 2003, I began publishing a series of books on hacks: *Google Hacks, Excel Hacks,* even *Mind Hacks.* One of the books we did in the Hacks series was *TiVo Hacks.* It wasn't a best-seller in the series, but the fact that people wanted to hack a consumer electronics product made me think that something was happening. If people were hacking TiVos without permission of the manufacturer, what was next? Would they start hacking their cars? Shouldn't every car have a "Preferences" menu? Should you be able to change the sound of your car horn? Shouldn't you be able to hack the doorbell in your home and in effect replace its ringtone? Why not look at things in the physical environment as if they were open to hacking?

How we started interacting with our computers was going to influence how we interacted in the physical world. We would develop the expectation that the physical environment should respond to us, change as a result of interaction with us, and in short be as adaptive as our software environments. Hacking wasn't limited to computers but was extending to cars, toys, watches, bikes, homes—almost anything you can think of. Hackers were hacking hardware, not just software. The physical world itself was becoming a play space, not just the rectangular LCD screen. We could hack the world around us.

BRIDGING DIY AND HACKING

I started *Make:* magazine in 2005 based on these observations. I originally intended to call the magazine *Hack,* but when I told my children about it, they didn't get it. I tried to explain that hacking was a clever way to solve a problem, but they weren't buying it. Instead, I decided to call the magazine *Make:*,[20] which was a word that could be understood by anyone.

I presented the idea to Tim O'Reilly in the back of a cab on the way to the OSCON (O'Reilly's Open Source Convention) in Portland, Oregon. I explained that this new magazine was "Martha Stewart for Geeks." It would be a magazine with "recipes" for projects that you could do, with new technology on the ingredients list. Tim found the idea interesting and encouraged me to pursue it, for which I am grateful.

I put a team together to create the magazine. For editor-in-chief I tapped Mark Frauenfelder of Boing Boing, which had started as a countercultural zine. He had worked as an editor for *Wired* in its early days. He's an avid DIYer himself. Mark understood on many levels what we were trying to do, mixing technology and DIY. He brought his own unique sensibility, as did our designer, David Albertson, who gave the magazine a clean, fresh design unlike a lot of technical magazines.

Before *Make:* there wasn't a contemporary magazine that reflected the DIY mindset around technology. Existing technology magazines viewed it in a narrow business-driven sense. They mostly covered the release of new products, but they didn't suggest satisfying projects for readers to do. I wanted the magazine and the projects in *Make:* to include all the technologies in our lives, not simply the newest: the old ones for cooking and woodworking alongside the new ones like 3-D printing and laser-cutting. Before *Make:*, there were DIY magazines for cooks and woodworkers but not for hackers. I set out to create a magazine for people engaged in personalizing, modifying, hacking, and creating, in the broadest possible sense. *Make:* is a bridge between the new world of hackers and the older world of traditional craftspeople, tinkerers, and hobbyists alongside the punks, crafters and DIYers. All these individuals share a DIY mindset, a determination to remake the world and adapt it to their own ideas, with the unstated assumption that this will make the world a better place. The magazine gave them a new name: makers.

I wrote in my opening welcome to the first issue: "More than mere consumers of technology, we are makers, adapting technology to our needs and integrating it into our lives. Some of us are born makers, and others like me, become makers, almost without realizing it."

The Internet has been a big driver to the rebirth of DIY culture and making. Websites have millions of user-submitted tutorials for anything from creating a Mad Hatter costume for Halloween to making a hula hoop from irrigation tubing. YouTube videos are now the way young people learn to do things that formerly might have required finding a class or having a person show you one-on-one. The Internet allows small producers to reach customers directly, and to succeed without corporate sponsorship or control. The online marketplace Etsy, launched in 2005, has grown to thirty million members and nearly one million stores. Kickstarter has provided a means to raise money to support the development and production of products designed by makers. Making can be found across a vibrant network of online communities with millions of passionate participants.

All of this activity—self-organized, collaborative, and widely distributed—is the Maker Movement, an agent of social change that includes all kinds of making and all kinds of makers, connecting to the past as well as changing how we look at the future. The Maker Movement is a renewal of some deeply held values, a recognition rooted in our biology, our history, and our culture, that making defines who we are.

2

Who: Amateurs, Enthusiasts, and Professionals

Today's makers can be hobbyists, tinkerers, artists, designers, inventors, engineers, crafters, and others. I resist saying a maker is one thing and not another. I believe the term *maker* resonates with so many because it's so inclusive and interdisciplinary. For our purposes, let's define a maker simply as someone who creates and shares projects. There are all kinds of makers. Some people make bread (full disclosure: I am one of them). Other people make airplanes. Some make sweaters. Others make robots. And some people make pumpkin-hurlers.

On the first weekend after Halloween, The World Championship Punkin Chunkin contest takes place in Millsboro, Delaware, in an enormous cornfield that's been bare since harvest. What started as a bar bet—who could hurl a pumpkin the farthest—has developed into an arms race, featuring air cannons mounted on the beds of semitrucks and a wide range of trebuchets, catapults, and hurlers. These days a nonprofit organization hosts the event, which has become so popular that it plans to move to the Dover International Speedway, having outgrown the cornfield.

I went to see it for myself in 2006 with Bill Gurstelle, author of *Backyard Ballistics*. The weekend was cold and clear. Large air cannons and trebuchets with patriotic team names such as Old Glory, Second Amendment, and Yankee Siege lined themselves up along the firing line, with over a hundred machines in a row, many of them flying American flags. One team was named Bad Hair Day, an all-woman crew in leathers.

The teams vary in size, as does their equipment, and they compete in different classes. Most are family and friends from the same town, and the event has become a ritual encampment. There is a lot of standing around waiting for something to happen, trying to stay warm, and naturally, there is a lot of beer. Meanwhile, kids from the different camps are running free and playing. Then, for a brief period, everyone gets serious. Each person on the team moves into position; someone shouts "ready," and someone yells: "Fire in the hole!" A small explosion of sorts occurs that blasts the pumpkin into the air and gets the crowd screaming.

In the event's early years they used leftover pumpkins. Today the Punkin Chunkin challengers use eight- to ten-pound white pumpkins grown specially for this purpose. They are a bit harder than a jack-o'-lantern and more like a gourd. When one of them explodes in midair, it is said to have "pied." Even on a clear day, it is quite difficult to see the pumpkin in air—it is just a tiny white speck that vanishes quickly against the autumn sky. Assuming the pumpkin didn't "pie," there's a flurry of activity at its landing spot out in the field, as a half-dozen four-wheelers circle the spot to measure how far it got. The pumpkins generally travel 1,000 to 3,500 feet, and a few exceed 5,000 feet.

One of my favorite competitors was a centrifugal hurler named "Bad to the Bone" from nearby Milton, Delaware. The hurler is mounted on a tower that sits on the bed of a truck. When it starts, the arms begin spinning slowly, round and round. As it picks up speed, the arms begin to blur together. Finally, channeling the energy of the arms spinning at top speed, with the truck and tower shaking and wobbling, the pumpkin is hurled into the air. It is exhilarating to see, although I made sure I was standing safely back.

The machines were impressive, but frankly, once I had seen one lineup in action, I became more interested in the people who built them. The Great Emancipator is an enormous air cannon painted bright red. Its team is dressed all in red, like a NASCAR crew with insignias on ball caps and over the breast pockets of their shirts. The team's leader is John Buchele, a tall man with a full beard from Jeffersonton, Virginia. I asked him casually how much Great Emancipator had cost him to build. He shrugged. "Over $70,000."

I wondered where he got the money to do this, yet I knew it wasn't about the money. This was a way to spend time together with the family, and have a sense of purpose and accomplishment. John and his team were very proud to be competing, and it was as much as about beating their own previous score as it was about beating anyone else.

I asked another man who had an air cannon mounted on a very large semitruck where he kept his machine when he wasn't at the competition. He replied without hesitating, "It's parked in my front yard." I wondered if he was married.

Then I talked to Steve Seigars of Yankee Siege from Greenfield, New Hampshire, leader of a nine-person team, most of whom were also named Seigars. Yankee Siege is a huge trebuchet made of iron. It weighs over fifty thousand pounds and stands sixty-one feet tall, not counting its throwing arm. Each of its four wheels are ten feet in diameter. It is amazing to stand next to Yankee Siege. It made me feel small. Yankee Siege won its class the year I was there, with a throw of 1,476 feet. The previous year it had won with a world-record throw of 1,702 feet. By 2013 they had surpassed the half-mile mark of 2,835 feet with a new machine, Yankee Siege II.

I had two questions for Steve Seigars. My first was, "Did you start off building small trebuchets before building one this large out of iron?" His answer was that Yankee Siege was the first one they built. They had never done this before. He and his family, who call themselves the Yankee Farmers, originally built the trebuchet as an attraction for their pumpkin farm.

My second question was, "What do you do for a living?" Steve replied that he was a dentist. I could not help but laugh. I said to him that he would probably not be remembered for his work as a dentist, but as the creator of the Yankee Siege.

Punkin Chunkin was a lot of fun, even more so for the participants. That's the whole point of competitions, hobbyists clubs, and associations, what I might call preexisting maker communities.

Another great example of this kind of community is the Experimental Aircraft Association (EAA). On several occasions I have visited the Oshkosh air show, AirVenture, held every summer in Wisconsin, which attracts more than one hundred thousand people. Oshkosh is a fly-in event, where pilots and their families arrive in their own planes and land at the adjacent airfield. They taxi to their campsites on a large field and set up tents under the wings of their planes.

EAA was formed after World War II, initially to lobby for the right to fly for recreational pilots, many of whom had flown in the war and were having trouble getting insurance to fly their own planes. Because of the difficulty in obtaining insurance, small airplane manufacturers went out of business. The EAA was able to get a federal law passed that said if you build at least fifty-one percent of your plane, then you can fly it.

One year Bill Gurstelle and I went to Oshkosh together. We met a pilot named Arnie Zimmerman who flew a two-seat open-cockpit airplane known by the name of its design: Breezy. It is the most minimalist idea of an airplane, closer to the Wright Brothers plane and a bicycle than to anything else at the air show. You can see how everything works because it has no exterior shell. The pilot sits in nothing sturdier than a deck chair with a cushion, positioned completely forward, a seatbelt the only thing keeping the pilot in the seat. The pilot has to wear goggles to keep the wind and bugs out of his or her eyes. Designed as a "family fun airplane," the Breezy design was developed by amateur aviators in the 1960s. The plans cost about $100 and the total bill of materials comes to about $15,000.

Arnie explained that he flew the plane up from Texas, dodging storm clouds as much as he could and setting down when he couldn't. He told us that commercial pilots like to fly the Breezy because they can really feel that they are flying. Arnie is the most casual sort of fellow, guffawing at the dangers of flying out in the open, and reminding me of Slim Pickens in *Dr. Strangelove*. It was clear he loved flying, and while talking to us, he was itching to get the plane back up in the air. If we wanted to keep talking to him, one of us would have to go up with him. I stepped back as Bill climbed into the seat behind the pilot and buckled in.

Bill later wrote of the experience: "At first terrified, I eventually got used to the feeling and the freedom. There's nothing between the flyer and the sky but a set of goggles. For a view and fresh air, the open cockpit that makes the Breezy so breezy—no door, no windshield—cannot be surpassed."

Because so many of its members build their planes, Oshkosh features a number of companies that sell kit planes. A kit plane can be as little as a set of instructions or a design for building a particular model plane. In that case, the builder needs to fabricate all the parts and assemble the plane from these instructions. A kit plane can also supply prefab components, which still require a lot of work to put together. Many of these kit-plane makers have a hangar facility where you go to build the plane, which, of course, typically doesn't fit in a garage. I met a father and daughter who had been going back and forth from their home in California to a facility in Texas in order to build their own plane. Eventually, once it was finished, they flew it home.

Building your own plane might be the ultimate DIY project. For many builders, the project might go on for several years, and there is even a market for selling half-finished planes. My brother-in-law, Rich Carlson, was one of those builders who had built the frame and done all the sheet-metal work in his barn but eventually sold the kit and its plans to another person who hoped to finish them.

I met Sebastien Heintz, president of the Zenith Aircraft Company, located in Mexico, Missouri, and one of the companies that sells kit planes. His father produced the original design for the Zenith planes, which has been revised over the years. Sebastien organized the "One Week Wonder" project at Oshkosh, where a team of people set out to build an entire airplane from one of the Zenith kits in one week. "We're doing this," he said, "to promote and to increase participation in kit aircraft building, and to demonstrate that building an entire airplane can be done by nearly anyone."

DEDICATED AMATEURS

Like many of the makers I meet, the participants at Punkin Chunkin and Oshkosh are dedicated amateurs, and proud of it. They form a community of interest, which really is a widely distributed group of people that share the same enthusiasm or passion.

The Latin root of the word *amateur* is *amare*, to love. Amateurs love what they do. For many, being an amateur is a kind of freedom to play without concern for making a living at it. Making can be seen as a hobby, a sideline, a pursuit outside the workplace. It's done on your own time. Saying you are an amateur doesn't necessarily mean that you are a novice, however. Amateurs can be highly skilled and committed, and the lines do blur between amateurs and professionals. Saying that you are an amateur speaks more to your goals than anything else.

Most of us are happy being amateurs and have no desire to become professionals. I like to grow and cook my own food. Cooking is something I've gotten good at, and in which I take pride. I have no desire to be a gourmet chef and work in a restaurant, yet I think as a cook I can appreciate what a chef does even more. I am content being a good cook who can create satisfying meals for friends and family. It's something I can share with others as I choose. Each of us is free to participate on our own terms.

I think of it as a pyramid of participation. At the base of the pyramid are the amateurs and at the top are the professionals. If we look at music as an example, we find at the base of the pyramid those who are learning to play an instrument as well as those who have played their whole life. Everybody has to start there, but many stay there. At the top of the pyramid are professionals, those who make a living playing music. Even there, only a few are superstars who command large audiences, compensation, and media attention. In the middle, in between the purely amateur and the professional, are those musicians who have gigs a couple of times a month or get paid to teach music.

Athletics fits the same structure, with a relatively small number of paid professional athletes on top and a much larger number of those who participate just because it is fun.

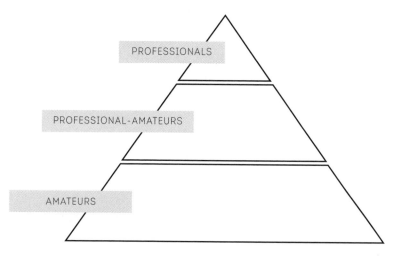

THE PYRAMID OF PARTICIPATION

A pyramid implies a hierarchy, and in fact, our culture tends to over-value the professional and undervalue the amateur. Amateurs can be seen as second-rate, objects of derision. "We use a variety of terms—many derogatory, none satisfactory—to describe what people do with their serious leisure time: nerds, geeks, anoraks, enthusiasts, hackers, men in their sheds,"[1] writes Charles Leadbeater, a British political adviser and author. For most of the last century, only professionals were taken seriously, and if we wanted to be taken seriously, we were advised to become professionals. In his 2004 essay "The Pro-Am Revolution," Leadbeater describes the rise of a new social hybrid he calls the pro-am, for professional-amateur: "Their activities are not adequately captured by the traditional definitions of work and leisure, professional and ama-teur, consumption and production."[2] The pro-am is in the middle of the pyramid.

We can use the pyramid of participation to bring attention to the large number of amateurs while noticing that there are professional and pro-am makers as well. Usually we think there's a natural progression from the bottom to the top. However, today the pyramid is expanding across its middle, as the bottom and the top meet: amateurs find more ways to get paid for something they love to do, and professionals find ways to do things they love to do as part of their job. Whether you call yourself a professional or amateur doesn't matter, nor do titles or credentials. What matters is what you do, and allowing that to speak for itself.

Alessandro Ranellucci is the Italian developer of the open-source Slic3r program, which is used to prepare design files for 3-D printers. He started developing the program in his spare time, and no one was paying him. He just saw the need for Slic3r and set off to write the code himself and give it away. At Maker Faire Rome, where he was an organizer of an area for makerspaces, Alessandro told me that Italian makers are seeking new alternatives for work that is meaningful to them. In a country where many say that the political and economic systems are broken, you wouldn't know that from meeting Italian makers who are filled with energy and enthusiasm. Alessandro said that some makers got involved because they didn't have jobs while others were quitting jobs, even taking less money to do something they loved. "Makers are choosing quality of life," he said.

If we characterize the Maker Movement as driven by amateurs, it's because makers will attempt to do things just to challenge themselves and take experiments about as far they can go, without asking anyone's permission or expecting a professional's compensation for their efforts. This is a kind of freedom that many professionals never experience for themselves, where permission and funding are often a prerequisite for action.

I believe in the power of the amateur to do something professionals won't or can't do, or plainly, just haven't done. I believe the novice can see things that the experts miss, and do things that the business-minded don't properly value. I also believe that ordinary people can develop uncommon insights and act on them because they have not been taught the "right" way to see things. It is a crazy and idealistic but ultimately democratic idea to believe that we can all contribute, and that art and innovation come from unexpected people and places. This is exactly what history shows is possible.

AMATEUR SCIENTISTS

The history of astronomy through the ages has proven the value of amateurs. William Herschel is famous for discovering the planet Uranus. Yet back in the mid-eighteenth century, had someone asked the young William Herschel about his profession, he would have answered that he was a musician. Herschel grew up in Hannover, Germany, and by the age of fourteen, he had developed a fascination with musical notation and the theory of harmony, and learned to play the oboe, violin, harpsichord, and guitar. At eighteen, his parents had him smuggled out of Germany, which was at war with France, and he ended up in London, where he did not speak the language, but earned a modest living by teaching music lessons and working as a church organist. Yet today we don't know William Herschel as a musician, because in February 1766, at age twenty-seven, he began keeping a journal of what he did at night. Night after night, all night long, Herschel stayed up gazing at the stars and recording his astronomical observations.

Herschel managed to acquire a small collection of both refractor and reflector telescopes. The refractor telescope, created by Galileo, was good for observing the moon and the known planets, but it was inadequate for looking much deeper into space. Newton had come up with a different type of telescope, known as a reflector, that contained a large mirror for gathering light, which enhanced one's ability to see dim objects that were far away. Herschel realized that he could improve the reflector telescope by using an even larger mirror made out of metal rather than glass. Because he could not afford to have one made for him, he decided that he would make it himself.

According to Richard Holmes in his book *The Age of Wonder*, Herschel by 1774 had "created an instrument of unparalleled light-gathering power and clarity."[3] He was the first to be able to see that the polestar or North Star was not one but two stars. Holmes writes: "By this means, Herschel began to build up an extraordinary, instinctive familiarity with the patterning of the night sky, which gradually enabled him to 'sight-read' it as a musician reads a score."[4] Herschel's significance, according to Holmes, is that he "began to conceive of deep space. He began to imagine a telescope which might plunge deep down into the sky and explore it like a great unplumbed ocean of stars."[5] One cool spring night in Bath, England, Herschel discovered that an object in the night sky had moved. It had been catalogued as a star, but Herschel's telescope allowed him to investigate further. At first he thought it might be a comet and notified the Royal Society, but later he asserted that it was the seventh planet, what would become known to us as Uranus. This professional musician and amateur astronomer made a discovery that fundamentally changed our understanding of our universe.

Forrest Mims, a well-known amateur scientist who wrote much-loved tutorial guides on electronics for Radio Shack, reiterated in *Make:* what he had written in *Science* in 1999: that some people think that amateurs can no longer contribute to modern science. He disputed the so-called "end of amateur science," saying, "Yes, modern science uses considerably more sophisticated methods and instruments than in the past. But so do we amateurs." He cited the ongoing discovery by amateurs of new dinosaur fossils, new species of plants, and new comets and asteroids. Forrest added that today's makers have "the technical skills and resources to devise scientific tools and instruments far more advanced than anything my generation of amateur scientists designed."[6] In addition, these amateurs are connected more easily to others, including professionals, and they can learn from each other and share their work.

Amateurs should be a key part of any healthy ecosystem. We can't have professional sports players without amateur sports. We can't have professional musicians without amateur musicians. We can't have professional makers without a broad base of amateurs who are doing this because they love it. To choose to do something for love, as an amateur does, to make something for no reason other than that we enjoy the process and to experience for ourselves the act of creation, is a powerful exercise of our freedom and a laudable demonstration of the human spirit.

I think of making, and even innovation, as a kind of sport, and my goal is to get as many people to play as possible. Amateurs play alongside professionals. Everyone requires the opportunity to play and needs practice to get good. If we get lots of people playing, we will have a better chance of discovering who has special talent, and even those of us who are less talented will benefit from being active and involved. I'm interested in untapped sources of innovation, untapped pools of talent. Many people have, for any number of reasons, never discovered their talent themselves, let alone have others discover it in them. With the Maker Movement as a framework for collaboration, making can be a team sport. Nobody has to go it alone.

SOCIAL CREATURES

Enthusiasm is the energy that amateurs burn. It keeps them up late at night, working on the weekends, staring at a problem much longer than most people until a solution is finally found, or iterating on an idea so many more times beyond what seems normal.

Though he uses the word "obsession," Kevin Kelly, contributing editor of *Wired* magazine, is talking about the force of enthusiasm when he said,

> Obsession is a tremendous force; real creativity comes when you're wasting time and when you're fooling around without a goal. That's often where real exploration and learning and new things come from. Even as a society we can have temporary obsessions with something that we will work through, and that's one way in which a society can explore an idea.[7]

Enthusiasts do what they do not just to please themselves, but also, importantly, to connect to others. Once you show interest in a hobby and begin talking about it, you'll find others who share the same interest. It's amazing serendipity. As we identify ourselves by our passion for our chosen activities and hobbies, we identify others, many of whom are better or further along in developing their interest. We join groups online and offline; we post photos of our projects; we develop Twitter followers. We not only share finished products with pride; increasingly, we are also sharing work in progress (or even a frustrating lack of progress).

One reason for this is obvious: we need help. To learn and improve, we need access to mentors or to people who just know more or can do more; these may be experts or peers. If I need to understand why the plastic filament in the 3-D printer isn't sticking to the print bed, I'm going to reach out to a person who can help me. Makers are usually generous with their time and they're willing to help, often because they recognize that others have helped them along the way.

Before the Internet, our opportunities to discover and develop a new skill were limited by our ability to find someone we knew who could explain it—in person. If you had a father who kept a workshop in the garage, you would have been exposed to tools and picked up on what he was able to do. The Internet is changing our ability to learn to make by improving access to other people who can help us. It doesn't matter where they live. We can learn from them by reading a blog or watching a video that explains how to do what they do.

Another important reason we seek out fellow enthusiasts is to stoke our own enthusiasm. To move from an idea to a project requires energy. If I have an idea, I am going to invest a lot of my own personal energy to pursue it. At first I have to find this energy in myself. Ideally, to be sustainable, my idea needs to generate energy, not just consume it. We often get this energy returned to us through interactions that our projects initiate. When someone shows that they enjoy my project—for example they share that they find it useful or amusing—they are giving me something back, and that can provide me with the motivation and energy to keep working. The currency of exchange is enthusiasm, something hard to define, but we know when we feel it.

In his book *Making Is Connecting*, British sociologist David Gauntlett writes:

> People often spend time creating things because they want to feel alive in the world, as participants rather than viewers, to be active and recognized within a community of interesting people. It is common that they wish to make their existence, their interests, and their personality more visible in the contexts that are significant to them, and they want this to be noticed….

There is also a desire to connect and communicate with others, and—especially online—to be an active participant in dialogues and communities.[8]

A big part of the DIY mindset is to seek out other enthusiasts who are doing similar things and to connect with them. Some people actually use the terms DIT, for "do it together," or DIWO, for "do it with others." Most makers don't conform to the cultural stereotype of the lone tinkerer who believes or fears that no one else shared his or her interest. They are social tinkerers, working in groups and in shared spaces. When we share what we are doing, we deepen our connection to our community, offline and online, physical and virtual.

CRAFTERS

In 2003, Leah Kramer in Boston created Craftster.org, a community site for "hipster crafters." Irreverent, edgy, dissatisfied with the notion of crafting as sentimental and crafters as stodgy, Kramer created an online community for crafters to meet each other and share their projects. The site's tagline, "No tea cozies without irony," is an expression of Craftster's subversive attitude toward traditional craft.

"I got bit hard by the crafting and DIY bug as a child, and it stayed with me throughout the years," said Leah. "As I entered my twenties, I got really bored of all the usual humdrum patterns and ideas you'd find in books and magazines at the time. I was really itching to make things that specifically expressed my own style, interests, and sense of humor. I remember I'd get excited when I'd see a cute little shop selling crafts only to walk in and be disheartened by the overwhelming smell of cinnamon and the sight of teddy bears dressed in handmade lace-trimmed aprons."

In 2001 a new holiday craft fair, the Bazaar Bizarre, had debuted in Boston, organized by a group of friends who were artists, musicians, and crafters. The first version took place at a VFW hall with a dozen craft vendors. "The things people were making were totally in the spirit of what I wanted to be doing with my crafty yearnings. Hand-knit puppets shaped like 1980s video-game characters, totally off-color cross-stitched towels." Leah joined the organizing team. The event expanded every year to include more vendors and take place in other cities.

That was offline. Online, there were a few communities in the early 2000s where Leah and other like-minded crafters could meet. "I loved visiting those sites, but it was really hard to share pictures, organize the craft ideas, and search for ideas." So when she started Craftster, she thought of it as "a place to share hip crafty ideas." Leah explained,

> Craftster was not only going to be a community I'd love to be a part of myself but also had this unique spin on things. The outside world seemed to think that the idea of young hip people who were subverting crafts with humor, irony, pop-culture themes was unique and new and worth talking about and writing about in various media outlets. That really helped Craftster succeed.

Leah had majored in computer science in college in the 1990s, but she had been interested in computers from childhood, fondly remembering her Apple IIe. She actually wrote computer programs to help her design new patterns for an intricate geometrical bead weaving that she liked. She recalls participating in Usenet newsgroups about her favorite bands and about crafting. "It's funny to think about the differences between then and now. If you wanted to share pictures of craft ideas and patterns with people, you had to print out pictures at the Fotomat and mail them via snail mail." She thought of the Craftster community as people who were itching to be creative in ways other than, or in addition to, what was already socially acceptable, like music (for example, being in a band) and writing and art (for example, creating your own zines). And at the same time they relished throwing people for a loop by doing the unexpected: crafting, something only grandmas did. These people felt empowered by taking crafting back and making it their own. And then, as the Internet grew in popularity and ease of use, these people got together and shared with each other what they were making and drew more and more people in.

She found people just like herself. "I enjoy having an idea and then building it out of practically nothing with my own two hands and brain. Sewing something out of nothing but fabric and thread has a similar feeling for me as making a computer program out of nothing but code.

"Craftster was definitely a side project I did just for fun. I had no idea where it was going to go and just wanted see what it would lead to. I wasn't even a Web programmer." Her side project grew and grew. At its peak, Craftster had 1.25 million unique visitors per month and 10 million page views per month. It became a valuable property. Kramer sold Craftster in 2010.

Suddenly there was a "new wave of craft" in America. In addition to websites like Craftster, craft was celebrated in Debbie Stoller's *BUST* magazine, as well as local groups like New York's Stitch 'n Bitch. Crafters in different cities organized their own brand of craft fair, from the Bazaar Bizarre to the Renegade Craft Fair, Craftin' Outlaws, Urban Craft Uprising, Shop 'n Mosh, Craft Mafias, and Reform School. Whereas handicrafts were once seen mostly as the domain of empty nesters, young adults under the age of thirty-five now dominate the $29 billion crafting industry.

In 2007 we launched *Craft* magazine as a companion to *Make:* with the tagline "transforming traditional craft." I saw *Craft* as a way to reconsider craft as a set of technologies and infuse it with new technologies, and as a new form of self-expression. The editor-in-chief was Carla Sinclair, who was married to *Make:* editor Mark Frauenfelder and an early partner in Boing Boing. She wrote:

> This DIY renaissance embraces crafts while pushing them beyond traditional boundaries, either through technology, irony, irreverence, and creative recycling, or by using innovating materials and processes.... The new craft movement encourages people to make things themselves rather than buy what thousands of others already own. It provides new venues for crafters to show and sell their wares, and it offers original, unusual, alternative, and better-made goods to consumers who choose not to fall in step with mainstream commerce.[9]

Projects in the first issue included making a stitched robot doll, a silver thread- and microprocessor-based programmable LED tank top, knit slouch boots, a minimalist "catnip castle," and an ant-farm room divider. *Craft* celebrated unconventional, unexpected, and unusual techniques, materials, and tools.

Crafter Manifesto

While studying for her PhD at the University of Helsinki in Finland, Ulla-Maaria Mutanen (now Engström) maintained a blog called HobbyPrincess.com. She came up with a manifesto that we published in *Make:*.[10] I found it to be one of the most insightful pieces on how to think about crafting and making:

1. People get satisfaction for being able to create/craft things because they can see themselves in the objects they make. This is not possible in purchased products.
2. The things that people have made themselves have magic powers. They have hidden meanings that other people can't see.
3. The things people make, they usually want to keep and update. Crafting is not against consumption. It is against throwing things away.
4. People seek recognition for the things they have made. Primarily it comes from their friends and family. This manifests as an economy of gifts.
5. People who believe they are producing genuinely cool things seek broader exposure for their products. This creates opportunities for alternative publishing channels.
6. Work inspires work. Seeing what other people have made generates new ideas and designs.

7. Essential for crafting are tools, which are accessible, portable, and easy to learn.
8. Materials become important. Knowledge of what they are made of and where to get them becomes essential.
9. Recipes become important. The ability to create and distribute interesting recipes becomes valuable.
10. Learning techniques brings people together. This creates online and offline communities of practice.
11. Craft-oriented people seek opportunities to discover interesting things and meet their makers. This creates marketplaces.
12. At the bottom, crafting is a form of play.

MEETING MAKERS AT A MAKER FAIRE

Because I enjoyed getting to know the makers I had met through the magazine, I began to think that they would enjoy meeting each other and talking shop. So in 2006 I came up with the idea for Maker Faire, and the first event took place in the San Francisco Bay Area. Cofounders Sherry Huss, Louise Glasgow, and I conceived of the Maker Faire as a reinvention of the county fair. The original county fairs, rooted in the agricultural economy, brought people from remote farms together to share their work—pigs and pies, as they say. "With these trading fairs came a cross-fertilizing of different cultures, and an imperative for joy and festivity," notes the National Fairground Archive on the history of the emergence of the county fair in England.[11] In short, fairs were a mix of exhibits, lectures, and a marketplace. They were a celebration of a way of life and its labors.

County fairs are still around, and I enjoy going to them and seeing the farm animals, the homemade jams, canned pickles, and tomatoes. However, many county fairs are in decline and frankly have gone stale, depending on aging rock bands and carnival rides to attract a crowd. Not many of us see ourselves as part of an agricultural economy anymore, despite growing interest in farmers markets, urban chicken co-ops, and backyard gardens. Instead, Maker Faire reflects our modern culture and economy. It is a hybrid of a science fair, a craft fair, an art fair, and a county fair. Maker Faire is an exhibition more than a competition. It's a people's World's Fair that is open to everyone who wants to show their creativity and technical prowess.

What Maker Faire does is bring together groups of people who don't otherwise get together—distinct subcultures, with different interests and different names. All of these groups exist side by side in our towns and communities, but they often see themselves as separate and not related to each other. Maker Faire helps them discover in each other the same kind of passion and sense of purpose.

Makers bring projects that range broadly across technology and science, craft and art. It blurs the lines between disciplines and interests. Mashing creative and technical work together makes it special. It's messy and noisy and thrilling. For one weekend each year, it is like the world's most creative, most agile, most innovative city. Unlike any other city, it is one that reflects our collective imagination and future potential. Its citizens are curious and adventurous—they love to learn and come to play. They are eager to move around, knowing that a surprise awaits them around every corner. Even if they stand in place, the wonders will find them, as so many are mobile: there are small speeding robots, spider robots that crawl, art cars that putter about, bicycles linked together like sled dogs, motorized Adirondack chairs moving alongside customized wheelchairs, and electric-powered cupcakes weaving in and out of the crowd.

The essence of Maker Faire is "show-and-tell": a context for makers to have conversations about what they do with other people, most of whom are strangers. The interactions are magical. We get to ask questions of makers such as: "Where did you get that idea?," "How did you get started?," "How do you find the materials you used?," "What was hard about the process?," and "Could *I* build it?" Our curiosity is amply rewarded.

At the outset, I didn't know if a broader public would find Maker Faire compelling. Even in our first year, Maker Faire attendance exceeded expectations, drawing eighteen thousand people. Maker Faire Bay Area, which was our first location, is in its eleventh year and now draws over one hundred thousand people, among them about 1,500 exhibiting makers. The flagship Maker Faires are a massive maker project themselves, produced by a dedicated team that loves the event.

Sherry, Louise, and I wanted Maker Faire to be, first of all, fun. We also wanted to attract families with children, not just adults. We wanted it to be a celebration of makers and making. Here is what I remember from our first Maker Faire:

- Acme Muffineering's electric cupcakes were developed by a Bay Area group for Burning Man. We had a baker's dozen speeding along the grounds.
- Bazaar Bizarre set up a craft faire with about forty hipster crafters.
- Douglas Repetto, the founder of Dorkbot, had a large artbot crawling on a wall to create a colorful drawing throughout the day.
- Make Play was a room filled with bins of recycled electronics and tables with soldering irons and hand tools. It was a free-for-all that only grew more chaotic as the day progressed.
- In the same area, master hacker Bunnie Huang was demonstrating breadboarding. A breadboard is a tool for prototyping the wiring of circuits.
- Wendy Tremayne's Swap-O-Rama-Rama was a clothing swap and remix. She organized designers and brought in sewing machines, so anyone could grab a few pieces of clothing and remake them into something personal. Scatha Allison of Jean Therapy, a designer of reconstructed denim, was there. A Fashion Show featured Swap-O-Rama-Rama models in remade designs.
- Limor Fried helped people build a Persistence of Vision (POV) toy kit, an electronic display that becomes readable as you wave it back and forth in front of your eyes.

- Lindsay Lawlor brought Russell the Electric Giraffe, a nineteen-foot-tall creature that rolled around the grounds and interacted with people.
- Steve Wozniak and a local team played Segway polo on a grassy field. Jamie Hyneman of *Mythbusters* was also seen playing.
- A homemade dump tank was a bucket mounted on a portable basketball stand; it dropped a bucket of water on a victim if one hit the target with a thrown ball. The Electronic Frontier Foundation sponsored it.
- The dump tank was next to a homemade barbecue pool heater by John Guy of Arizona. By the end of the day, it was heating the water for the dump tank.
- Jon Sarriugarte and Kyrsten Mate from Oakland brought SS Alpha Fox, which looked like an official NASA space vehicle and shot balls of fire in the air.
- The Crucible, an Oakland space for industrial arts, brought its fire truck, which sprayed fire instead of water.
- A maker who called himself Crabfu, whom we later got to know as a game designer named I-Wei Huang, brought beautiful steam-powered robots.
- Lenore Edman and Windell Oskay, a maker couple from Sunnyvale, California, who would later start Evil Mad Scientist Laboratories as a business, brought an interactive LED table that they had built and which was designed to respond to motion. "The sensors in our table were going crazy because of the number of people," Lenore remembers.

Without our even realizing it, a community had been born. Many of the makers met each other for the first time but would continue working together for many years. The community members were the cocreators of the event. We just gave them space, but they brought the creativity, passion, and ingenuity. Makers were the stars of the show, the talent.

Their projects represent work that is still at a raw, early stage. Each exhibit represents a kind of experiment.

Above all, Maker Faire provided makers with valuable feedback. A reaction such as "Cool!" can mean a lot. The Faire also gave makers the chance to make new connections and meet interesting people they might have never met on their own.

Maker Faire is contagious. All kinds of people comment that Maker Faire has this "vibe," a feeling that is hard to explain but easy to experience for yourself and see in the expressions on people's faces. It's the feeling that anything is possible, as we revel in experiencing the creativity and talent in our community. Everybody brings their best self to Maker Faire, and indeed, they are creating something together that none of us could do by ourselves. Everyone is happy; there is the sense that we all "play well together." It makes us feel optimistic about our society and our future. That this spirit continues to flourish after more than ten years is truly incredible, but I can't say I'm surprised.

Maker Faire has served as a catalyst for the Maker Movement, inviting more and more people to participate and to see themselves as producers, not just consumers.

Lessons from Maker Faire

- We are hungry to learn, particularly from each other.
- We are fascinated by, not intimidated by, intensity, depth, and complexity.
- We want most to express our own creativity, which leads us to recognize it in others. And vice versa.
- We want to make connections to people in our own local community as well as the Internet.
- We can acquire the skills and knowledge to get better at doing almost anything.

3

What: Art, Interaction, and Innovation

What we were looking for in *Make:* magazine was cool projects, much like vintage editions of *Popular Science* and *Popular Mechanics. Make:* was to be filled with DIY projects, just like magazines for cooks, gardeners, knitters, or woodworkers, with descriptions of the materials needed and detailed instructions as well as some explanation of how and why a project works. We would include small, medium, and large projects: some might take an hour, others might take a weekend, and some might require several weeks of work. One difference from the other magazines was that we'd embrace new tools and technology.

For the very first issue, our team needed to choose the quintessential *Make:* project to feature as our cover project. It had to be an exciting project that conveyed what we were about and that couldn't likely be found in any other magazine. My team and I considered many fun, fascinating projects.

It was the Silly Putty that really grabbed our attention. The project was about kite aerial photography (KAP), "step-by-step instructions for building a very low-cost rig consisting of a camera cradle made of craft (Popsicle) sticks and model airplane plywood, a shutter-button timer mechanism that uses rubber bands and Silly Putty, and a camera-stabilizing suspension." The project involved building a cradle for a camera that would be suspended from the kite, and then figuring out how to snap a photograph once the kite was in the air.

This KAP project became the first ever cover for the magazine: a mash-up of flying a kite and shooting pictures with a camera. Both the project and the magazine were later included in an exhibit at Cooper Hewitt in New York City in December 2006 when *Make:* was selected by the Smithsonian Design Museum for the National Design Triennial, a show meant to reflect current trends and ideas in design. We had an actual kite aerial photography rig hanging from the ceiling above the magazine.

The maker was then a professor of architecture at the University of California, Berkeley, Charles C. Benton, known as Cris. To show new perspectives of buildings for his students, Cris had begun developing his own camera rigs to take pictures of buildings, seeing them from a bird's-eye view. This was in the days before drones were commonly available, and Cris was seeking a perspective that could not be obtained either from ground level or from a helicopter or an airplane. Kite aerial photography places us "in this region just above our heads and maybe up to two hundred feet," said Cris. "It can be very difficult to occupy this space with a helicopter or with an airplane. This lower-level view no one has much explored, and there are not many techniques that will put you there. With a kite you can hang a camera there, and sort of hover. You are close. You can see detail." Cris designed a low-cost rig with a Silly Putty viscous timer that he describes in a how-to video:[1]

> There is a little brass tube into which is stuffed Silly Putty, and then another tube is stuffed into the Silly Putty. I take a rubber band and tie it down and loop it around a pin. That pin will move around ever so slowly. Basically the rubber band is waging war against the Silly Putty, but the Silly Putty ultimately loses. When the pin rotates all the way around to facing upward, which might take two minutes, then the rubber band will slip off. Then a lever goes up, and another rubber band pulls another lever down toward the camera, where there's a little nubbin that pushes the shutter button. When the rubber band goes relaxed, the shutter is actuated, but it also drops the ping pong ball as a visual indicator that the shot has been taken.

It's a genius Rube Goldberg sequence, just to take a single shot from the sky. To take another picture with the apparatus, he had to bring the kite down, advance the film in the camera, reset the viscous timer, and then get the kite back up in the air.

KAP is not new, although Cris' low-cost rig is innovative. KAP has its own history and an active online community. A famous photograph of San Francisco three weeks after the 1906 earthquake was taken using several kites to lift a forty-nine-pound camera one thousand feet in the air.[2] Online, there is a KAP community with a variety of clubs that you can join, such as the Japan Kite Aerial Photography Association. Many of these clubs have cloth badges for members to use on their gear to identify their allegiance. The KAP community shares new designs and improvements on existing designs as well as their own experiences in practicing KAP in different locations and contexts.

"One of the neat things about kite photography is that you have to invent your own gear," said Cris. "I'm always tinkering around at my workbench building apparatus. How can I achieve getting a camera in the air and pointing in the right direction, and sufficiently stable to capture a useful image?" Cris's workshop contains all of the different rigs he's built over the years. The Popsicle-stick cradle with a disposable camera and a Silly Putty viscous timer rig was designed to be low-cost and easy enough to build so that students in a class could construct one. For his own purposes, however, Cris uses more advanced rigs built to hold a heavier SLR camera in a metal cradle with servomotors that he can remotely control along three axes. He uses a "picovet" suspension system to keep the camera level, using parts he harvested from an old hard drive.

Cris learned a lot about kites as well, even making some of his own. He often traveled to a shoot with his trunk full of different kites. Choosing the right one for the wind conditions meant that the camera could get up in the air and be stable.

While Cris's initial interest in KAP was related to architecture and photographing buildings from the air, KAP became a hobby for him once he realized that a kite-suspended camera gave him unique views of many things, including the coastline around the Bay Area. Eventually, he began exhibiting his KAP photographs in galleries, where the perspective and beauty of his photos were appreciated by others as art, apart from the techniques used to create them. In 2013 he published a book called *Saltscapes* that features the abstract compositions of colorful salt ponds found in the southernmost part of the San Francisco Bay.[3]

In *Saltscapes,* Cris reflects that his twenty-year involvement in kite aerial photography went through three phases. "The first phase," he writes, "focused on the mechanics of KAP—learning about kites, refining camera cradles, studying urban aerodynamics, and developing strategies for camera survival…. The second phase was accumulating lots of experience taking photographs in different places under varying conditions." He began to focus more on the composition of his photos, thinking through the best way to take a shot. In his third phase, KAP "became a means to explore and document specific landscapes."[4] It became less about mastering technique and more about applying the techniques more purposefully.

Over time Cris evolved from a KAP enthusiast to an aerial photographer with an artist's vision. In his case, all of it taken together might be seen as a life's project, a dedicated effort that can define a person as much as a career.

PROJECTS, PROCESS, AND PRODUCTS

We wanted projects in the magazine to be fun, a bit clever, and somewhat unusual. That's what we meant by "cool." Here is a list of some of those projects:

- Make a cigar-box guitar. Add a piezoelectric pickup and connect it to an amplifier and speaker embedded in a cracker box.
- Use barbell weights to build a $14 camera stabilizer to allow you to hold steady a video camera on the move.
- Build two tiny solar-powered BEAM robots, one that looks like a satellite and one that looks like a scooter.
- Repurpose an old VCR and use its timer to build an automatic cat feeder.

- Make your own soda bottle rocket and shoot it off in your backyard.
- Make a cubicle-detection system so you might know if anyone invaded your space while you were gone.

From the outset, our goal was to find projects that other people could replicate and that people would have fun doing. However, we didn't believe that many people did most of the projects in the magazine, but that they could learn from all of them. What we hoped people would do eventually is come up with their own projects.

A project is how makers organize and share their work. The project is the working out of an idea, making it real. A project involves a list of materials and tools. It becomes something that you can talk about and share.

A project represents a step-by-step process, a series of actions in sequence. It is also an iterative process, as it often requires doing it more than once to get it right. Some people call it a design-build process. There are formal ways to look at this process, such as design thinking. Anytime we go through trial and error and iteration, that's a design process, whether it's formal or informal. The important thing is that the process becomes yours, and you learn about that process and improve it from one project to the next.

A process is like a narrative or an adventure: it begins with an initial intention, the stated goal of the quest, however rough or well-formed, but there are unexpected challenges and misunderstandings along the way before the end is reached. Documenting a process over time can seem like a tedious task, but doing so allows a maker to reflect, gain insights, return to old methods, or embrace new ones. Sharing the documentation allows others to learn from the experience.

Launched in 2005, Instructables.com is a place to publish step-by-step instructions for any kind of project and reach a global audience. The site was created by a group of super-smart MIT graduate students who shared a passion for kite-surfing, and upon getting their PhDs, formed Squid Labs and moved to the Bay Area. Squid Labs was established as an innovation factory that would generate new projects and spin them out when they discovered a commercial opportunity. They rented a warehouse and filled it with the kind of tools that they had been able to access at MIT. Eric Wilhelm was a member of the group, and he revealed how the idea for Instructables actually developed from their passion for kite-surfing:

> The combination of sailing, unpredictable weather, experimentation, and sheer power strongly appealed to each of us. The sport was still its infancy, and the gear was unrefined and way too expensive for anyone on a graduate student budget, so we built our own. We'd turn up at beaches around Boston with hand-sewn kites and boards shaped from plywood. Half the equipment would break and the other half would perform beautifully. We'd then document our results on our personal websites and a blog called Zeroprestige (which has since moved to the Instructables ZeroPrestige group). Soon we were getting e-mails from people asking for more information, wanting to meet us at the beach, and looking for tips on finding and building similar communities. As a result of freely sharing our work, we met a ton of great people … and were smacked in the face with the need for a web-based documentation system.[5]

Instructables follows a typical maker project pattern in which the sharing of information creates community and opportunity. Eric eventually spun out Instructables on its own, took on investors, and became its CEO. I remember visiting Eric at his office and seeing that he had four pairs of eyeglasses on his desk. He said he was retraining his eyes, using different corrections to stimulate his eye muscles, so that he wouldn't need glasses. "In a year I've gained one diopter," he told me. He's a wonderful self-experimenter.

Christy Canida, who was Eric's classmate at MIT and later his wife, had been working in biotech and supporting them while there was little revenue coming in from Squid Labs. She eventually joined Instructables herself to focus on building the community. Christy established a "Be Nice" policy for the online community that asks members to be "positive and constructive." Nothing can discourage people from sharing creative work more quickly than experiencing negative or dismissive comments.

Instructables, which was bought by Autodesk in 2011, has become an extensive repository of more than sixty-five thousand "recipes," and a demonstration of how people feed off each other's creative work.

INTERACTION

Some makers are identified with a single project, such as Lindsay Lawlor and his nineteen-foot-tall giraffe named Russell. Others move from one project to the next, covering a range from artistic to absurd to entrepreneurial. Marque Cornblatt is one of the serial makers. At the first Maker Faire Bay Area, Marque brought his video robot, named Sparky. The first Sparky was all analog, an assemblage of junkyard parts attached to a motorized wheelchair. It used old-fashioned pirate-radio transmitters and receivers for video. Marque said that the video quality was "like flying through a snowstorm." He improved Sparky, transforming it into a digital telepresence robot. "As a maker, I'd throw out last year's technology and start with new technology," he said. Sparky went from three hundred pounds down to about six pounds. As Marque considered what to do with Sparky, whether it had commercial potential or not, he eventually concluded that he was just done with it, and he open-sourced the designs.

Marque's craziest creation came next. He called his project Waterboy, but it was also known as Buckethead. "For Burning Man, I wanted to come up with something that was as absurd as possible for the deep desert." He wanted to seal himself in a suit filled with water, kind of the opposite of what a diver bell is. "I connected to people who had professional expertise in special effects or building wet suits, and I said, 'Here's what I'm trying to make.' They told me, 'No you can't do it. You're going to die.'"

He realized that what he was trying to make was a lot like a waterbed. He found a Bay Area waterbed manufacturer who generously agreed to make the suit for him. I saw him walking around Maker Faire, and he looked like a man who took an oversized goldfish bowl and stuck it on his head—with the water still in it. "With Waterboy I felt like I was a test pilot," he said. Marque teamed up with the band OK Go at Maker Faire, and they went on stage in the Waterboy suits, with Damian Kulash singing a love song underwater, a truly remarkable performance.

Marque said he has never been completely comfortable pursuing art. He grew up in Baltimore and went to NYU, majoring in film and television production. In high school, he worked on sets for the theater department, which was his first exposure to using tools. "In New York, I was doing everything but school," he said. He was introduced to Adam Savage, later of *Mythbusters,* and together they set up an artists' cooperative workspace and then a tiny storefront gallery called Points of Departure. He and Adam shared a workbench, which "is a great way to get to know someone," he said. After Marque moved to California and got a job building theatrical effects, he encouraged Adam to join him, and they became roommates.

He came to drones through remote-controlled cars and planes. He started getting together with Justin Gray in Oakland to "smash our toys together." Eventually, this became a weekly gathering they called Fight Club, and they began crashing drones on purpose.

"The first thing we learned from setting out to crash drones was that commercial drones were super fragile and the parts were expensive," he said. "As a beginning pilot, you crash a lot and do a lot of damage to your drone." He wanted to figure out how to make the airframe for drones more rugged. He launched a Kickstarter project promising "to build an airframe that didn't need to be repaired." Marque knew that he could produce a video that would not only help him raise the money but also gain momentum for gathering community around what was first called Game of Drones and then Aerial Sports League. After all, it's a lot less fun to fly your drones alone.

In the game, drones do battle inside cages, trying to knock out their opponents, colliding in midair and often falling to the ground. If a drone cannot be repaired quickly by its pilot and returned to flight, it is out of the game. Because some drone pilots were more interested in acrobatics than battles, he began adding competitions where pilots could demonstrate new tricks. Aerial Sports League competitions now include racing with drone pilots wearing first-person view (FPV) goggles. They are racing a new breed of drone—"tiny, fast, and angry like hornets," said Marque. He believes that the lower cost and level of expertise will help drone racing become more accessible for entry-level racers.

A competition is an important way of pushing the limits of a technology. Henry Ford was one of the early organizers of auto racing, staging a number of match races and attempting to set land speed records. His goal was the kind of publicity that would make automobiles popular. He wanted people to talk about what cars could do, and he saw making faster cars as a unique design challenge. When Ford started racing, nobody thought of themselves as race car drivers. Ford had to recruit a competitive cyclist, Barney Oldfield, who was completely unfamiliar with the controls of a car. Going over a one-mile track cleared on a frozen lake, Oldfield was the first person to drive a car at sixty miles per hour. It was a Ford, and racing helped make a fortune for Henry Ford.

Marque wants people to get excited about drones, and he thinks that racing will do that. Events like Aerial Sports League are creating a new category of competition, much like the X Games did for skateboards and BMX bikes, developing a distinctive language of tricks and maneuvers, and promoting broader interest in drones by more people, even among those who don't fly them. "Today, my creative energy wants to be applied to creating a sustainable business," said Marque, who now sees performance drones as his opportunity.

Like the Aerial Sports League, many of the projects I see at Maker Faire are designed for interacting with people. Nightmare Kitty was Phoenix Perry's project, which I first saw in her Brooklyn, New York, gallery. It was a game for kids, but she also viewed it as an interactive art installation and a computer science project. In 2011 she brought Nightmare Kitty to the World Maker Faire in New York. Her project provided a fun, interactive experience, but perhaps it has more going on than the player realizes. To participate in Nightmare Kitty, the player stands in front of a screen on which the video is projected. Round icons representing kitties fall from top to bottom, and the player tries to avoid having them fall on them, ducking and dodging to get out of the way.

Phoenix Perry grew up as a poor and gifted child in the South. She thought of herself as a freak, and her parents worried that she'd burn down the kitchen if they didn't find something for her to do. She loved playing with a construction toy set named Capsela, which was created in Japan by the Mitsubishi Pencil Company. She got an Atari computer, on which she first experienced video games. She taught herself to program and eventually got a job in game development in Silicon Valley. It was a good opportunity, but the demands of the job were too much. "My hands shorted out," she told me. "So many hours of coding at really old terminals." It took her about five years to recover, during which she began learning about neurophysiology and how the brain is wired.

Nightmare Kitty was designed based on the research of Amy Cuddy, who wrote about power poses.[6] "Neurotransmitters change based on the position of your body," said Phoenix. "The goal of Nightmare Kitty was to put kids in a low-power position to kind of trigger their stress hormones." They would then feel relief when they stood tall.

Phoenix was one of the first to use the Microsoft Kinect to create interactions based on how the player's body moved. "You shouldn't need to type or use a mouse to interact with a computer," she said. "Kids encounter scary things. They can feel helpless, powerless, and confused." She wanted them "hunkering down to avoid falling kitties," because "it changes your body chemistry. When they crush the kitty, and defeat what scared them, they feel really powerful and safe. They feel it in their body."

Phoenix said she likes that artists have begun taking over technology. "There's no need to present programming outside the act of being creative. I see computing as a kind of poetry—inside the framework of creative practice and play." Phoenix now teaches physical computing at Goldsmith University in London. She also created the Code Liberation Foundation to teach game development to young women.

Maker projects are creative applications for new and old technologies, combining mechanical, electronic, and digital systems. Some projects are serious and others are silly. Some help us connect with other people in new and interesting ways. Other projects can turn into a useful commercial product, something that can be produced and sold.

MAKER TO MARKET

Lisa Qiu Fetterman did not think of herself as an inventor, per se, nor an entrepreneur. She did not know about makers, and she did not grow up considering herself to be one. She loved food and cooking, and that passion led to her to develop a new product and eventually start a company.

Lisa Qiu came to the United States from China at the age of seven, and her family settled on Long Island, New York. She explained:

> At school, I was basically a pariah because I didn't know you were supposed to change your clothes every day. In China, we changed clothes every week. When I finally realized what the problem was, it was too late—I had developed a reputation. Nobody wanted to come to my house. And then, one day, somebody brave decided to come over, and we were super excited. My dad got dressed in a suit. My family said: "Lisa, it's your first friend."
>
> We served them a really expensive delicacy, a thousand-year-old egg. The problem was that the egg is, well, putrid; it's green and smells like Limburger cheese. You've got to be really into it to enjoy it. Once my friend took a bite of it, she didn't speak for the rest of the night. She went home. The next day I went to school really scared. Before school, I considered asking my parents if we could move.
>
> But at school, everybody came up and asked me if they could come eat that weird food at my house. Yeah! So from a really young age, I realized, 'oh, people connect through food.' I've been obsessed with food ever since. When I got into NYU, I went to Babbo, which was the closest restaurant to school, run by Mario Batali. I walked in and begged him for a job in Italian, and he gave me one on the spot. I worked in the kitchen and in the front of the house, wherever they needed me. If you get paid $8 an hour, people will basically let you do anything if you show active interest.

Her parents wanted to know what she would do with a humanities degree. Her triple major at NYU, in journalism, metropolitan studies, and American studies, was how she showed her parents that she was willing to work as hard as anyone. After graduation, she kept working in restaurants until she got a job at Hearst Corporation's digital department.

She met her future husband, Abe Fetterman, who had moved to New York City after finishing his PhD in astrophysics at Princeton, at a fancy gym. "On our first date, we talked about food. He said he was really into food, and I said, 'oh, really?'" She told Abe about sous-vide cooking:

> I wanted to save up money to buy a $1,000 sous-vide machine, because it's the best way to cook. In every single restaurant that I worked in, we had one: a huge hulking piece of laboratory equipment that we relied on for seventy percent of our components. I started thinking this has to be in everybody's home because it's so easy to use. All you do is put your food in a vacuum bag and drop it in the water and walk away. That's it. It's crazy easy. Abe said, "Let's just make one."

Essentially, a sous-vide cooking machine functions by immersing food inside a sealed bag in water that is kept at a constant temperature—much lower than normally used for cooking—over a long period of time, three to five hours. Lisa and Abe's first homebrew version used a bowl to hold the water and an immersive tea-heating coil to control the temperature. It sort of worked.

Abe did some research and found a DIY sous-vide project by Scott Heimendinger in *Make:*.[7] "It required soldering, which we didn't know how to do," Abe recalls. "We didn't know how to do a lot of things," adds Lisa, laughing. "But we knew that it was possible to DIY something. Abe said he could do it." That was a promising sign as far as she was concerned.

The essential element of a sous-vide cooker is the PID controller. PID is the proportional integral-derivative (PID) algorithm. In *Make:*, Heimendinger explains what this device does:

A PID temperature controller is like an advanced thermostat. Regular thermostats click on and off when the measured temperature passes fixed thresholds, and the resulting temperature oscillates around the thresholds. This creates an opportunity for temperature "carry-over" in the food, like the way meat's internal temperature can rise after you take it out of the oven. In contrast, PID controllers use the PID algorithm, common in industrial control applications, to track temperature changes and calculate how much ON time will raise the measured temperature (the process variable) to the target (the set point) asymptotically, rather than overshooting it. By predicting, mathematically, the impact that turning on the heaters for one second will have on overall water temperature, the PID controller enables extremely precise temperature control and stability over long periods of time.[8]

One of the places Abe and Lisa met regularly was a vegan restaurant, the World Café, near NYU. "You can sit upstairs for hours and they won't bother you," said Lisa. That's what they were doing when they overheard a conversation at another table about making and makers. A writer for *Make:*, Matt Metts, was interviewing Mitch Altman, a well-known hacker and maker of the TV-B-Gone device that allows you to turn off any TV by remote control. Altman travels around the world teaching soldering classes and advising hackerspaces. He just happened to be in the same café as Lisa and Abe that day.

As Lisa heard them talking, she said to Abe, "What's a maker? It seems like we might be makers." Not shy, Lisa went up to them and started asking questions. "Mitch gave me the key to his hackerspace and invited me to his soldering class in Brooklyn." It was literally an unfinished basement in someone's house. After the soldering class, they found an Arduino class at a Brooklyn hackerspace named NYC Resistor and decided to take it. I asked Lisa and Abe if they had known what Arduino was before that. "No," said Lisa. Abe knew how to program in C++, and he found Arduino easy to use. From what they learned, they began building their own sous-vide cooker, controlling the electronics with an Arduino.

Over months, Lisa and Abe came up with the design for a DIY sous-vide kit. They named it Ember to signify that it uses a gentle heat. The Ember kit had a set of instructions and a list of required parts. It was an open-source kit; they shared all the information needed to build the cooker. Abe said that it probably took him took eight or so hours to put one together. It wasn't easy, but it could be done—and for a lot less than $1,000.

Lisa and Abe had solved the problem that they set out to address: Lisa had her own cheap and portable sous-vide cooker. By doing additional research, she realized that she wasn't alone in wanting one. "I searched on Twitter for #sousvide, and I talked to everybody on there. I asked them what they knew about it. I wanted to know if they would be interested in a DIY sous-vide cooker." She specifically looked for New Yorkers and asked if they'd like to get together.

She met a cheesemaker, Yoav Perry, who was interested in how sous-vide could be used to control temperature during cheese-making. Eventually she visited him at his house. "For the cost of materials, we put together a DIY sous-vide for him. Yoav was someone we found randomly on Twitter, and now he's the official cheese guy for all of Mario Batali's Eatalys." Slowly but steadily, she built a network of others who shared her interest and enthusiasm.

In 2011, Lisa and Abe decided to set up a demonstration of their DIY sous-vide cooker at Maker Faire in Queens, New York. They wanted people not just to see the cooker but to experience the food that had been cooked in it. Lisa decided to make eggs, perhaps recalling how her first friendship at school was forged. They made a thousand sous-vide eggs the night before Maker Faire, and then battered, breaded, and deep-fried them—so they were "crunchy on the outside and melty on the inside"—on site at Maker Faire. They ran out of eggs well before the show was over. People were thrilled. Lisa and Abe were even featured in a NPR story about Maker Faire. In response to all the excitement they created a website, QandAbe.com.

"People bought our kit, not just dudes, but older women," said Lisa. They also held classes at NYC Resistor to help people make the DIY sous-vide cooker themselves. "It would take their entire day, but they were happy," recalls Lisa. They continued to think about taking the next step in the development of a finished consumer product. They weren't sure how to move that idea forward.

Lisa heard about HAX, a hardware incubator that was starting up in Shenzhen, China. An incubator, sometimes called an accelerator, works with start-ups in return for a piece of equity in the business. They coach its founders, introduce them to a network of mentors, and generally advise the start-up on how to develop a marketable product. HAX was the first such incubator in Shenzhen that promised to help maker start-ups navigate the manufacturing ecosystem in China.

Lisa applied to join and was accepted, but at first she and Abe decided not to go. They both had jobs. Abe's day job had brought them to San Francisco, where he worked as a physicist on compressed-air batteries. "At the restaurant my work schedule was 4 p.m. to 2 a.m.," said Lisa. "Abe's work schedule was 9 to 5, so we never saw each other." Abe remembers telling Lisa, "We're engaged to be married, but we never see each other." They changed their minds and decided to go. The move to China, Lisa thought, would be like a pre-honeymoon. It turned out to be a lot of work.

As part of the deal to be part of HAX, Lisa and Abe agreed to HAX taking a seven percent equity stake in the company and $15,000 up front in funding. "We didn't care about seven percent equity," said Lisa. "People think that's a lot, but we thought of this as our weekend project." In Shenzhen, Lisa and Abe worked out of Seeed Studio, a maker business started by Eric Pan. "Our story is like finding out that Madonna hung out with Andy Warhol, and Steve Jobs gave John Lennon one of the first Apple Is. There were all these connections we made, and they helped us so much along the way."

Their goal in the HAX program was to develop a working prototype. "We learned about sourcing. We went to visit factories all around the area. Thankfully I speak Chinese," said Lisa. She also helped as translator for many of the other teams in HAX who were not from China.

In the middle of their time at HAX, the two ran out of money: "Straight up, all gone," said Lisa. They had no savings and were $20,000 in debt. Lisa believes that is why they felt so free to do what they did, because there was nothing left to lose. "I was so upset. I came up with a brilliant idea: to sell my kidney. You can get as much as $18,000 for a kidney." Abe said no to her crazy idea and instead suggested that they go on vacation. They decided on Bangkok because it was close and they knew someone there. At NYC Resistor, they had met a young Thai chef named Wipop Suppipat. He had been one of the students in their DIY Sous Vide workshop, and now he was living in Bangkok.

"Before we left on the plane, I e-mailed him everything we were working on. He picked us up in the morning for breakfast and he said to us, 'Did you know that I have an industrial design degree from the Rhode Island School of Design?' We spent our entire vacation working with him, and he flew back with us to China to be our cofounder. Together we finished our prototype."

The prototype was a long gray cylinder with a clip that allowed the device to attach to any pot. On top of the cylinder was a green knob. You press the center of the knob to turn on the device, and turn the knob to set the temperature. Nobody at HAX really liked Ember as a product name, so Lisa and Abe made up a new one. "We definitely wanted 'nom' in it," she said. "Because when you're eating, *nom nom nom* means it's so good." In Chinese *nom* means "water" and *iku* means "eat"; Nomiku was born. They later learned that *nomiku* means "drinking and eating" in Japanese.

Of her experience at HAX in Shenzhen, Lisa said that "in the end it was all about meeting other makers, and at first, you don't know that." Lisa and Abe went back to the United States to get married in Brooklyn. They used the Nomiku prototype to cook perfect steaks at 135 degrees Fahrenheit for their wedding guests.

The next challenge was to raise money that would allow them to take the prototype into production. Lisa knew about Kickstarter, a crowd-funding platform, and realized they could ask the community of sous-vide enthusiasts to back their project. "I begged our wedding videographer to help us make a Kickstarter video." Nomiku's Kickstarter video featured Lisa demonstrating the Nomiku, saying something like, "I'm not a top chef, but now I can cook like one." It's warm, friendly, and personal. It also contains a few scenes from the Qiu-Fetterman wedding, and she explained the secret to the perfect medium-rare steaks her wedding guests enjoyed: the Nomiku.

The Kickstarter campaign for Nomiku was launched with a goal of $200,000. In thirty days, they had raised almost $586,000 from 1,880 backers and became the number-one most funded project at the time in the food category. The couple returned to China to find a factory and get the product made. Lisa also learned that she was three months' pregnant.

Nomiku was a prototype, but it wasn't designed for manufacturing (DFM) ready. "First, we had to try to get a design that we could make. That's a huge challenge. We had to go back and forth with the factory and work together with them. They didn't really have the resources to pay attention to us. When you first tour a factory, you see all the amazing things that they make. You think: that's the factory for us. Then you realize they make those amazing things for clients who pay them millions and millions of dollars." Lisa shook her head. "Oh, I get it. I don't have that kind of money. They're not going to pay any attention to me. Oh, this is bad," she said, in a different voice that seemed to mock her. "That's what happened: you find out that your job is not a priority. So then we knew that we basically had to do it ourselves."

They had to do several different phases of verification testing: First, to verify that the engineering works after making several samples; then to make sure that the process of putting the parts together works; finally, to look at the production assembly line and determine whether it can be made easier to assemble. It took about a year to get production units for Nomiku, probably faster than the average manufacturing time for most Kickstarter projects, but too long for Kickstarter backers who are waiting for a product they already paid for.

"People were really mad at us for a long time, but then they were really happy when we shipped. David Lang, author of *Zero to Maker*, likes to say everything's forgiven when you ship, and for the most part it was. We were also very open about what happened, and we took lots of pictures and updated our backers every two weeks." After two years in China, Lisa and Abe came back and settled in San Francisco. They had a product to sell.

Lisa didn't have a marketing budget or a sales staff; she had to do all that herself, and still does. "I'm out and about. I'm going to restaurants talking to them about it; I'm going to food and wine shows; I'm going to Maker Faire. I'm on the Internet forums; I'm going on TV if I can get it; I'm going on the radio if I can get it. I'm contacting food journalists who have written about sous-vide. I'm hosting my own parties."

Mostly, Nomiku was sold directly on their website and through Amazon. They did not have the margin in the product to sell it retail. "We had a fancy meeting at Williams-Sonoma headquarters. We dressed up and talked to the head of appliance buying. It was pretty cool. He said, 'You know, we take fifty-five percent margin.' And we said, 'Oh, we can't have that.' We would be losing money on each unit."

Then she began to see competition. She had priced Nomiku at $299, and now her competitors were releasing a product for $199. Many of them also used Kickstarter and raised even more money than she had. "They looked exactly the same as ours," said Lisa. "It was discouraging, but I had mixed emotions. First, I thought that anyone who's telling me sous-vide is not going to be ubiquitous is an idiot, because look at these other examples. If I go into a business meeting and I tell you my market is huge, would you believe me if I didn't have these aggressive competitors?"

It validated what she believed to be true, but she realized that she'd just spent two years or more on a project that she felt she was being ripped off. "I reminded myself that this is what I've always wanted to do. I just felt like it was too early to be discouraged. I don't really blame people for copying it. If it was such a passionate fire in my mind, there must have been somebody else who was thinking about it." Lisa eventually figured out how to differentiate Nomiku from the competition on terms other than price:

> First of all, we're the smallest, the most powerful, and easiest to use. And once we shipped Nomikus everywhere, we got a flood of feedback on what people wanted in their machines. We took the ideas that kept coming over and over again and made changes based on them until we came up with a whole new

machine. That's the Wi-Fi Nomiku. The smartphone app lets me be a sous-chef in your kitchen and help you use your sous-vide machine, because it turns out people have lots of questions about temperature and time.

The Wi-Fi Nomiku became the next-generation product. They launched a second Kickstarter in 2014 and raised $750,000 from 5,438 backers. Another advantage that Nomiku had was its sizeable community, thanks to Lisa's networking.

After their exasperating experience with manufacturing the first Nomiku in China, basically doing everything themselves and having to keep close watch on every step of the manufacturing process, Lisa and Abe were determined to try making the Wi-Fi Nomiku in the United States. "We're already crazy for doing all this stuff. This just adds another layer of crazy."

They talked to Jen McCabe, who had founded a company called Factorli, a small-batch manufacturing operation in Las Vegas, with venture backing from Tony Hsieh of Zappos.com. Lisa had not visited Factorli but felt confident they could work with Jen and signed a contract to produce twenty-five thousand Nomiku units. However, the day before Lisa was to launch the Kickstarter for the Wi-Fi Nomiku, Factorli abruptly closed. "We had the Kickstarter video with Factorli in it," said Lisa. "We had to frantically edit our video to take it out."

Abe began researching how to set up their own production line. While he was in China at the factory, Abe had seen opportunities to do things more efficiently:

Once we needed a specific washer, and we were told that nobody made that washer within two hours of our factory. So we took a boat to Hong Kong, found a shop that had the washer, bought it, and brought it back on a boat, because there was no more efficient way to get that washer. That's a ridiculous thing to have to do in order to get your product made. If I was in the United States, I could go on the Internet, order it, and have the part delivered overnight.

One thing I took away from China is that people there are really good at making things that they have already made, but you have to be the expert in making your product. So in the end we now rely on China to make things like our pump heaters. They make millions of them. But nobody makes Nomiku. By doing it on our own, we have more opportunity to take advantage of new processes and new ideas.

Abe also wanted the engineers and sales teams to be in the same space so they could work together closely. He had seen how MakerBot had its engineering in China and sales team back in Brooklyn, and the two teams had trouble syncing up, each group blaming the other for delays. The most expensive part in Nomiku is the printed circuit board: its electronic brain. They ended up having that made in China. The plastic casing and other parts required tooling for injection molding. For that they ordered a tool that would be made in Taiwan and then shipped to their San Francisco office.

At the end of 2015, Nomiku started with five people on the assembly line. Abe said that it was hard finding people in San Francisco to work in production. "People aren't really used to doing this in San Francisco. Turnover is high, or else you have to pay a lot."

After many years of work, Lisa's passion and purpose is still strong. Nomiku, which employs ten engineers and designers in addition to the newest assembly line workers, is about to ship a maker-made product in a competitive new category. I asked Lisa if others should follow a path similar to hers. "Everybody should—absolutely everybody," she replied. "It's so much fun. Yes, it's painful, but there's so much purpose. If I were doing something that I didn't believe in, I would be the most miserable person in the world. Instead, I'm stressed out but incredibly fulfilled."

Lisa's story demonstrates the power of the Maker Movement to engage people with a passion who end up building not just a product but a community with which they share that passion. Lisa was able to take advantage of an emerging ecosystem that includes incubators, crowd-funding, and marketplaces. The maker-to-market ecosystem is still developing, which is why people like Lisa and Abe are really the pioneers, having to figure things out the hard way. Yet the fact that it even exists enables more people to develop new ideas and test them in the market.

MIT Sloan School professor Eric von Hippel explores user-center innovation in his book *Democratizing Innovation.* It is users, not manufacturers, who drive innovation and create not just new products but also new niche markets. Ordinary users like Lisa and Abe are motivated to create the things they need, things they can't get from manufacturers. Von Hippel gives the example of users in extreme sports. A person wanted a kayak for running the rapids in rivers; it had to be shorter than existing kayaks. Unable to buy one, some users made it. Others saw what they had made and wanted one. What started out as something that supported a hobby turned into a business.

Von Hippel writes that when users innovate, they are the chief beneficiaries of the innovation. Yet it turns out that when you solve your own problem, you find others with the same problem who can use your solution. According to von Hippel, user-centered innovation typically involves open sharing of ideas and designs, as well as the development of what he calls "innovation communities."[9]

Makers start out a project with a goal of satisfying themselves, and they keep working on it until they do. Even if they don't set out to start a business, sometimes their project develops into one.

4

Where: Communities, Schools, and Industry

While making can and does take place almost anywhere—on the kitchen table, in the yard, in the basement—the serious tinkerer, hobbyist, or crafter has always needed a dedicated space: a "room of one's own," in author Virginia Woolf's words. Men traditionally took over garages or, as the English and Aussies like to call them, sheds: usually a rough, unfinished, messy space that was both a workshop and a haven apart from family and office. Women traditionally had craft rooms or project rooms in the home. These spaces had a dual purpose: to set aside a place for special projects, and to organize the materials and tools needed for those projects.

Carl Bass, the CEO of Autodesk, is one of those fortunate people who has his own dedicated workshop, neither at his office nor at his home. It is certainly not modest or humble, even though there is something of those qualities in Carl himself. A mathematician by training, Carl has always worked with machines to pursue his own interests, which range from practical to artistic. He is probably one of the few CEOs who could call himself a maker. On a tour of his workshop in Berkeley, California, he showed me a collection of baseball bats that he had made for his son's Little League team. He took out a blank and started making another one, rather effortlessly, peeling away the wood in strips.

"What I'm finding interesting is the mix between traditional tools and more computer-controlled tools," he told me, moving from a 1950s-era drab-green lathe to a tall, all-white, five-axis Thermwood CNC router that towered over him. He referred to this industrial machine as a "beast." It is not the kind of machine you normally find in someone's hobby workshop, and Carl's workshop is not something that's available to most people. In cities, space itself is expensive, let alone acquiring all the equipment. Instead of private workshops like this one, the Maker Movement is seeing the rise of community spaces that support the sharing of tools in a common space. The workshop and its tools are being democratized: made more accessible for more people.

These community spaces go by a number of names, including hackerspaces; fabrication laboratories, aka fab labs; and TechShops, with *makerspace* a generic and inclusive term that I use for all of them, whether they are nonprofit or for-profit and based in schools, libraries, universities, or corporate campuses. In the end, what it is called doesn't matter. The purpose of the space and the community of makers that comes together there are what's important.

Travis Good, a former AOL executive and trained as an engineer, visited more than one hundred of these spaces in the United States over two to three years starting in 2011. Travis likes to go on road trips, so he would get in his car and visit as many spaces as he could in a particular region. "I was fascinated by these spaces, who was building them, and how they worked," he told me. He described the genesis of many such spaces as a group of people meeting each other, and "someone would say, 'We should get a space where we can make stuff.'" If the group moved forward, they had to accomplish a number of tasks, which Travis said included everything from finding an affordable space and negotiating a lease, getting permits, ordering equipment and supplies, and most importantly, finding other people in the community to appreciate the need for the space. Most of the peculiar features of any space are simply due to circumstance: "Hack Manhattan could only afford a small room in New York City, but Dayton Diode could cheaply lease a large space." The leadership of spaces seems to vary, including, according to Travis, "dictator, democracy, or anarchy." However, Travis was struck more by similarities than differences.

I was most happy to see the emergence of consistency among successful makerspaces. Facilities tended to have clean and dirty labs as well as distinct areas for learning, working, and social-izing. They had a governance structure that survived the blows of change while still serving their communities. Membership tended to be tiered, recognizing that some can pay more while others need a break. Education was a consistent theme that played out in classes, in projects, and in group builds. Most did outreach, raised funds, hosted events, and welcomed new makers readily into their community.

In this chapter, we will look a variety of makerspaces. There are five different types of organizations:
- organic local spaces started by a small founding team
- community spaces that create an organization, usually a nonprofit
- for-profit membership spaces organized as a business to provide access to tools and training
- collaborative development labs funded by institutions such as colleges or companies
- museums, schools, and libraries that intentionally create public spaces for making

There are many benefits to having a social atmosphere where you can find someone who can share the expertise you want, where you can be inspired by the projects and processes of others, and where you can make new friends. Meeting other people who share your interests is key not only to your satisfaction but also your development as a maker. It signifies the shift I previously described, from the lone tinkerer who worked alone in a shed to the social tinkerer who thrives in a space shared by others. If the mythical garage involved in the start of Ford, Hewlett-Packard, or Apple is any indication, the makerspace will be the new place where future inventions are incubated.

Makerspaces play an important role in expanding who has the oppor-tunity to become a maker by expanding access to the tools and expertise required to make things. I have called them the "on-ramp" to the Maker Movement, attracting new people by providing training and creating a supportive environment where more people can learn to do something that they might have thought was too hard for them.

Makerspaces also are a "commons" that can play a key role in a greater economic transformation. As economic adviser Jeremy Rifkin describes in his book *The Zero Marginal Cost Society,* the twenty-first century is seeing networks mostly replacing markets, social capital being more important than financial capital, and the ownership of goods becoming less important than access to those goods. "Millions of prosumers are freely collaborating in social commons, creating new IT and software, new forms of entertainment, new learning tools, new media outlets, new green energies, [and] new 3-D-printed manufactured products,"[1] he writes, describing the "collaborative commons."

Of course, we've seen community-based spaces providing access to resources as far back as guild halls, and then in the form of libraries, playgrounds, and community centers; a makerspace is just the latest iteration of the commons.

HEATSYNC LABS

Hackerspaces started out as informal, small, predominantly male clubs that appeal to the hard-core geek, focused more on computing than making. Even so, many hackerspaces would not be happy with that definition as they have changed, adding tools for making and opening up to the broader community.

However, a number of hackerspaces that I've visited still seem like clubs. A small group runs the space, and they have might have a dozen members, with people bringing in tools and resources from home to share with the group. Some feel like an eccentric's garage full of scavenged treasure, waiting for someone to whip it into shape. Some are good about collecting members' dues, but others are dependent on last-minute pleas for the generosity of everyone involved to make rent. They tend to serve the needs of their founders or core members without having staff to manage the facility.

Just one of hundreds of possible examples, HeatSync Labs occupies a small storefront on Main Street in Mesa, Arizona. It opened in 2009, claiming to be the first such space in Arizona. HeatSync Labs describes itself as a "grassroots co-op of volunteers." The space is longer than it is wide, with worktables down the center and workbenches on the sides. Since they are on Main Street, people will wander in, curious to know what goes on there.

HeatSync Labs hosts study groups and meetings for young makers and crafters, among several examples of community outreach. Its members teach classes and host demo nights where they show off their projects. A recent show-and-tell was on the subject of model rocketry.

While HeatSync Labs encourages membership, the space is open for free to anyone. Like most hackerspaces, it is most active on nights and weekends. Each member of HeatSync Labs has access to a storage box to use to put things away, but as anyone who's ever had a roommate knows, some people are just not good at picking up after themselves. A makerspace can begin to look pretty cluttered if everyone leaves projects out in a shared workspace.

To keep the clutter under control, HeatSync has a system of issuing "parking violations" to members who leave their projects out on a workbench and don't clean up after themselves. One can apply for a "parking permit" to a leave a project and its materials on a table overnight. I learned that the ticketing system had actually been developed at Twin Cities Maker's Hack Factory in Minneapolis by Christopher Odegard, another example of how good ideas are propagated in the maker community.

On a visit to HeatSync Labs, I caught up with two young teenage makers, Schuyler St. Leger and Joey Hudy. I had met both through Maker Faire. Schuyler is known for having given one of the best-ever introductory talks on 3-D printing, which I had seen on YouTube. He's smart and inquisitive, and can speak on any technical subject as though he has memorized the Wikipedia entry. Joey is a young maker icon, having been invited to the White House to represent Maker Faire. Joey demonstrated his orange marshmallow cannon to President Obama, firing a marshmallow across the room, generating an astonished look on the president's face.

Jim St. Leger, Schuyler's dad and an engineer at Intel, said HeatSync was a very engaging community for his son, and even though everyone was eight to twelve years older than him, they considered him a peer. "I can recall whiteboard sessions with Jasper Nance, currently an engineer at Orbital Sciences, where she would answer endless streams of questions from Schuyler," said Jim. "She would end up teaching him all sorts of engineering and physics equations, math equations and concepts, and so much more." Jim recalls that Joey Hudy learned to do his first printed circuit board at HeatSync Labs with help from its founder, Jake Rosenthal.

HeatSync Labs is only one of many different collaborative working spaces in the Phoenix area. At least seven other community makerspaces exist, including one for teens at the public library. TechShop in Chandler is a fully equipped fifteen-thousand-square-foot space developed in partnership with Arizona State University (whose students can use the facility for free). Also in Chandler is Local Motors, a company that produces and sells kits for off-road vehicles and has a facility for building them. The number of collaborative community spaces in many major cities is growing. Interestingly, new spaces don't seem to be regarded as competitive with existing spaces.

ARTISAN'S ASYLUM

When I first walked into Artisan's Asylum, I was greeted by a person behind a reception desk. I walked into an open sitting area with old couches and comfortable chairs where members were hanging out with each other. From that vantage point, I could see the machine shop, the electronics workbenches, and the woodworking areas. I could also see several rows of stalls, an important feature of the space that I learned were rented out to local artisans. I was surprised by how large and open the space was, and I thought it must be the largest if not the best community makerspace in the country.

Artisan's Asylum is located in Somerville, Massachusetts, in a working-class neighborhood that borders Cambridge. With support from the city, Artisan's Asylum took on a forty-thousand-square-foot space, formerly the Ames Envelope factory, which closed in the 1980s. Half of its membership lives within a mile's radius.

Founder Gui Cavalcanti graduated from Olin College in robotics, an engineering college in Needham, Massachusetts. He worked for a while for a defense contractor doing robotics but was driven by his desire to have access to the kind of facilities he had at college, the "always-open workshops and tight-knit creative community." Molly Rubenstein came on to manage the space and become executive director, which eventually became a paid position. Molly had a background in the arts and in nonprofit management. Both Gui and Molly shared the goal of developing a community space, and they had complementary skill sets: Gui focused on organizing the shop layout and equipment, while Molly built the nonprofit organization.

Gui's first attempt at a makerspace failed, but he applied what he learned from that experience to Artisan's Asylum. The main lesson was that it's hard to make ends meet without ongoing support from the community. Artisan's Asylum was started with a $40,000 budget for opening the space and outfitting it. Most of the tools were used, either donated by members or acquired for the cost of removing them from a former worksite. By not having the money to buy everything that was needed, Gui involved the community in the creation of the space, which resulted in its membership being even more invested. Artisan's Asylum looks unfinished, and undoubtedly that is intentional because its members are in the constant process of changing it and adapting it. It may never be done.

Gui and Molly recognized that in a city with high rents and small apartments, artisans lacked dedicated space, and many had to pursue their craft on nights and weekends. So Artisan's Asylum is like a village of makers, crafters, and artisans. There's a density of creative and technical talent that tends to attract more people to join. Bringing all these people together creates opportunities to learn from one other. Many of them teach workshops to the community. Gui said, "the original goal of the space was to democratize the act of making something from scratch."

Artisan's Asylum is a model makerspace, based on the scope of the space, its membership base, and the overall sense of community it fosters.

Gui and Molly also began documenting their model and sharing it with others. I worked with them to organize a workshop called "How to Make a Makerspace" in 2013 that offered information on working with real-estate developers, city administrators, insurance brokers, and even lawyers. The event sold out, with several hundred people coming to learn about how to start a makerspace in their city or town. Gui gave them a packet of materials that he used, including a very detailed spreadsheet that explained their budget. Artisan's Asylum represents a collaborative model for a well-managed makerspace that is an asset to its community, and others want to copy it. Even though Gui and Molly moved on to other endeavors in 2015, Artisan's Asylum has continued to thrive.

DALLAS MAKERSPACE

Located in an industrial park, Dallas Makerspace developed out of a robotics hobbyist club that met monthly. My guide when I visited was Doug Paradis. With his son, he had been involved in the original robotics club. An engineer who had retired in 2009 from Texas Instruments, Doug developed the Tiny Wanderer robotics kit for use in schools. I visited the makerspace for the second time in 2013, and since my first visit in 2011, they had doubled the amount of space they had, and memberships had grown from 80 to 240.

I visited on a weekday night, and the place was busy. As I looked down the main corridor, I saw multiple rooms on each side, one for electronics, one for crafts, and another with couches for hanging out. Another room was a lab for synthetic biology projects: Doug showed me how they were growing a tissue culture to make a very thin fabric that one of the makers would use to sew into a dress. At the end of the hall we opened the door to a roughly finished space: part woodworking room, part auto garage. The CNC machines on one side of the room were silent, as were the welders on the other side. In the middle was a black motorcycle with a whiteboard propped against it that named its owner and his project: "Honda ST1100 Electric Conversion." Taped to the board was an electrical diagram.

Doug introduced me to Michael Eber, one of the most active and enthusiastic members. Eber suffers from medical conditions that cause discoloration of his skin and swelling, yet his spirit is undiminished. Some of his projects, like an interface for Arduino to your cell phone, are commercial. Others are more whimsical, like the vending machine he acquired and connected to the Internet so that members could order components for projects online and get them immediately through the vending machine. Instead of candy and chips, the Dallas Makerspace's vending machine serves breadboards, Arduinos, packs of small electronic components, wire, and batteries.

Michael wasn't the only person tinkering with vending machines. Another member had designed a game interface for a vending machine. It had a video display and four colorful buttons. When I approached the machine, it prompted me to take a quiz. If I answered the question right, it asked if I wanted to cash out or continue to play. The more questions you get right, the better the prize you get—unless you get it wrong, and then you lose whatever you gained.

Doug Paradis also introduced me to a nineteen-year-old woman named Kirsten who was there with her parents. After high school, Kirsten turned pro as a tennis player but had lost her ranking after a car accident. During her recovery, she checked out the makerspace, and now she was learning to make robots, with Doug as her mentor. Someone remarked that Kirsten was the best solderer in the place, and she smiled bashfully. I learned that she had been hired to work part-time building custom circuit boards for another member's commercial project.

It is always great to see a makerspace with women members. I asked several women I met there what makerspaces could do to welcome women and increase their involvement in making. One was Stacy Devino, an engineer who, with her husband, created a modular programmable LED display system called LEDgoes, which successfully raised more than $17,000 on Kickstarter. Stacy said what she wished for was not that anybody do anything for women specifically, but that people would just stop turning women away from engineering. She personally had the experience of counselors, family friends, and professors offer advice that she might prefer something other than engineering. They assumed that they knew what was best for her because she was a woman, Stacy said, even though they hardly knew her.

Three makerspaces opened their doors in 2013 and 2014 with the mission to serve only women or those who identify as female: Seattle Attic in Seattle; Flux in Portland, Oregon; and Double Union in San Francisco.

Georgia Guthrie, executive director of The Hacktory in Philadelphia, established a goal of inclusivity when she came on. The Hacktory has reached a fifty-fifty gender balance among volunteers and organizers. Georgia wrote in an article on the *Make:* website that men might assume that the lack of women in makerspaces is due to a lack of interest among women in making things, but that's not right. Many women have had negative experiences around technology, with people underestimating their ability or even being dismissive. "When a woman walks through your door," she asks, "is the general assumption that she must be a beginner or that she's tagging along with someone else? Such assumptions may be based in real experiences, but to address this problem, lay these experiences aside."[2]

Georgia believes that each space should evaluate the biases in the organization and among its members, and then decide what it can do to make everyone feel at ease and completely able to explore any activities they like. It's clear that we need to work on making makerspaces more welcoming and inclusive, especially to women, people of color, and people with disabilities.

FAB LABS

Fab labs, short for fabrication laboratories, originated at MIT, the brainchild of physicist and professor Neil Gershenfeld, after he realized how little experience students had with physical machines—with atoms as opposed to bits. He created a legendary class, "How to Make Almost Anything," open to anyone at MIT. The class filled up quickly each semester with students from a variety of fields—architecture, design, art, and engineering—anxious to learn about the range of machines and tools.

The popularity of that class led to the permanent creation of the first fab lab as part of the Center for Bits and Atoms at MIT's Media Lab. The program is a reflection of MIT, which Neil once described to me as a "safe place for strange people." Occupying an entire floor, it is many makers' wildest dream come true: access to all the machines you could dream of, without much supervision. I can imagine that its lack of structure would be overwhelming for some, yet it seems to be seen as freedom by most at MIT. The motto of fab lab is "Learn, Make, Share."

In Neil's incredible vision, humans will be able to build self-replicating machines: machines that can not only build other machines but build new versions of themselves. By extension, he wanted to build labs that would build themselves. While the vision of self-replication hasn't materialized yet, fab labs have nonetheless been widely replicated. Having a network of labs all over the world that can make anything is a powerful idea.

The fab lab model established the standard for a shared workshop and mirrored what Neil's students could access at MIT: CNC machines, milling machines, drill presses, 3-D printers, laser cutters, and water jets. By standardizing the list of hardware and software in all fab labs, a maker can design in one location, prototype in another, and iterate in yet another, able to count on the availability of certain tools and machines.

In the first issue of *Make:*, Neil laid out his vision, which was also the subject of his 2005 book, *Fab:*

> The idea that inspires us is that the next revolution is going to be the personalization of manufacturing: using accessible digital technology and machine tools to program the physical world we live in, just as we today program the bits in worlds of information.... By personal fabrication, I mean a desktop machine that can create three-dimensional structures as well as logic, sensing, actuation, and display.[3]

For that article and several times since then, I visited the Center for Bits and Atoms at the MIT Media Lab, an academic R&D center that has been at the forefront of research for digital fabrication. Neil has his own kind of DIY mentality. He seems determined to build everything himself (or with grad students), because the things he envisions or desires don't exist, or what already exists is completely inadequate. A rapid thinker and a tireless leader, Neil is the engine behind the Fab Lab Network, spreading these spaces around the developed and developing world.

The closest fab lab to MIT is across the Charles River in South Boston. When I visited the South Boston Fab Lab, which is run by Mel King, a civil-rights leader in Boston, a group of teenagers were using a screen printer to make T-shirts. All around them were high-tech machines, but what captured their interest was using a screen printer to apply ink on a design of their own creation on a cotton T-shirt. I asked Mel about it. He told me, "When those kids go out in their neighborhood, wearing their T-shirts, and someone asks where they bought it, they can say, 'I made it myself.'" I smiled in appreciation.

Today, there are over six hundred fab labs around the world and in a variety of different settings, such as a school of architecture (Barcelona), a science museum (Chicago), a community college (Lorraine, Ohio), and a public storefront (Tulsa, Oklahoma). Neil believes that one day the best and brightest students won't have to come to MIT, at least physically. Via fab labs, with access to a common set of tools and processes, students will be able to share knowledge and designs, and learn and collaborate across international borders.

FACLAB

FacLab was the second official fab lab in France, started by Laurent Ricard and Emmanuelle Roux at the University of Cergy-Pontoise, north of Paris in Gennevilliers. When I visited the space, in a new building with stark white walls, it had been open fifteen months. Still, there was plenty of evidence of work in progress. Laurent, who gave me a tour, apologized for the mess. He needn't have done so—it was tidier than most makerspaces I visit.

I saw a student using a high-end laser cutter for an architecture project. A class on using the laser cutter was held later by the local guru, a teenager named Ilyes. On a newly arrived ShopBot desktop model, someone had cut a beautiful piece and left it unattended, so I couldn't ask questions about it. In a textiles room, I saw a fascinating project in process, recreating a historical costume using modern tools. There were several models of 3-D printers, an older MakerBot and an Ultimaker among them. I also met a maker named Julien Desprez working on the design of his own 3-D printer, which goes by the name DOOD, for "digital object on demand." He's since run a successful crowd-funding campaign on KissKissBankBank, a European alternative to Kickstarter, and is excited about making and selling his DOODs.

FacLab is open to anyone. The only thing asked of people who use FacLab is that they document and share what they do. I really connected to their vision: to promote the growth and development of individuals, and to encourage self-directed learning, experimentation, and sharing.

"It is not a technical place; it is a social place," Laurent told me, admitting as well that he was disappointed with people's reluctance to come in and try things out if they were unaffiliated with the university. "People have a hard time believing that this facility is open to them, that anyone can come in and use it without paying anything. It's an open bar, and they can't believe it. Or they might suspect we have a hidden agenda." Nonetheless, FacLab has people who drive a considerable distance to come to use the facility.

We had a brief conversation about terms: I wanted to know if people identified with the English words *maker* or if there was a French equivalent. Laurent mentioned *bricoleur*. But *bricolage*, sometimes also used in English, has connotations of folk art or pastiche: "construction or creation of a work from a diverse range of things that happen to be available." We also talked about *bidouille*, which is more like tinkering and hacking but is usually used in reference to children's activities.

In a conversation I had in Paris at a tiny makerspace called Le Petit Fablab (not part of MIT's network, and not open to the public, but where a company called Nod-A decided to share the fabrication equipment they use in their business with selected inventors), a researcher named Véronique Ronin commented that "*fab lab* is the fashionable term for spaces dedicated to making" in Europe. It is preferred over *hackerspace* or *makerspace*, even when it is unaffiliated with the MIT network.

Paris has had two Maker Faires, and the word *faire* in French is the verb "to make." In February, Jean-Louis Missika, deputy mayor of Paris, announced a plan for Paris to become a "Cité des Makers." It appears the word *maker* is becoming a part of the French language.

FAB CAFÉ

There are Fab Cafés in a number of cities. The first opened in Tokyo in 2012 as a place where you could use 3-D printers and other technologies while you enjoyed a coffee or tea. You didn't have to be a member; you could just pay for the time using the machine. Like any café, it was a social space.

Cecilia Tham opened the second official Fab Café in Barcelona. Tall and intense, Cecilia is a confluence of Asia, America, and Europe. She was born in Hong Kong, lived in Macao as a child, and then moved with her parents to Atlanta. She went to college in Boston (Harvard, she admitted when pressed). There she met her partner, an architect from Barcelona, where they moved in 2006 and soon had a child. She struggled, she told me, trying to figure out what she should do. On his home turf, her partner was able to get a job, but she wasn't. "He bought me a sewing machine, and I almost killed him," she said. "But I started using it," she admitted, emphasizing, "I mean, every day."

Good things started to happen. "Every time I made something, my self-esteem grew a bit. I had turned an idea into a physical thing. For the first time in my life, I did something on my own—not my parents, not the government." Then she built a shutter-release remote for a photographer friend, and started to think she could make anything. It was a maker epiphany.

On a trip to Tokyo, Cecilia walked into the Fab Café and realized she wanted one. "If I had that need, I thought that others would too. So I built it." She found a storefront in Barcelona and began to design it. She added a coworking space, called Makers of Barcelona (MOB). She also created a consulting service called MEAT—Make Extremely Awesome Things.

It took a while for the Fab Café to work in Barcelona. After the first year she broke even, but she believes the café is sustainable. It serves one audience that she didn't expect: retirees. "They come in and drink coffee, and aren't actually fabricating anything," she tells me. "But it's a place where they can come and learn about technology. They've asked me for workshops." Cecilia is proud that she made this happen, on her own terms and on her own.

TECHSHOP

TechShop is a chain of commercial makerspaces, the first of which opened in an industrial park in Menlo Park, California, in October 2006. When he showed up in a vintage military transport vehicle at the first Maker Faire in April 2006, hoping to gain interest and support for the idea, founder Jim Newton pitched it to me as "like a gym where you get a membership to use the equipment." His dream was to have unlimited access to the tools of a machine shop, plus new tools for digital fabrication such as laser cutters and 3-D printers, perhaps $2 million worth of equipment, for a relatively modest membership fee that hovers around $100 per month.

Currently TechShop has more than ten locations in the United States, three of them in the San Francisco Bay Area, and others in Los Angeles; St. Louis; Austin; Arlington, Virginia; Dearborn, Michigan; Pittsburgh; and Chandler, Arizona. They recently opened the first TechShop in Europe in Paris and will soon open a location in Abu Dhabi in the United Arab Emirates. More are on the way.

TechShop's downtown San Francisco location on Howard Street is the full realization of what a TechShop can be. Its three-story layout is clean and open, with plenty of windows, a large central workspace, and separate rooms for electronics, 3-D printing, woodworking, and metalworking. It can be busy at almost any time, day or night: they recently announced twenty-four-hour access seven days a week.

What happens at a TechShop is probably true of most makerspaces. On the one hand, there is a group of makers who show up with a pretty clear idea of what they want to do. Often their project has some practical or commercial application, and they need reliable access to a workspace and tools to make it real. However, the majority of people who show up at TechShop want to belong but don't have a project or a purpose. They want to learn how to use the tools, and maybe that will help them come up with a project.

TechShop has a good record of attracting people who already have ideas about what they want to build and just need tools and access. Having to pay to use a TechShop usually means its members are more serious about why they go, and many of them are trying to create a new product or launch a company. Yet I find just as important the story of David Lang, who wrote the book *Zero to Maker,* in his experience walking into a TechShop, knowing little about it beforehand, and deciding that making was something he wanted to learn to do. He wanted to leave behind a job in a cubicle, develop a whole new set of capabilities, and explore new opportunities. David eventually met Eric Stackpole, and the two of them founded a company called OpenROV to build underwater robots.

Mark Hatch joined Jim Newton to lead TechShop after working at Kinko's corporate office, seeing a similarity between the chains: roomfuls of copier machines and roomfuls of fabrication machines. A fast-talker with spiked white hair and a good sense of humor, Mark said that he is focused on raising capital to open sixty to a hundred TechShop locations by 2020, with five hundred members per location. This is quite a goal: it's hard work to run one space effectively, but it's even more difficult running them in different geographic locations. A large, well-known for-profit makerspace in Brooklyn called Third Ward declared bankruptcy after overinvesting in expansion. This is the challenge that TechShop is trying to meet, and it's not easy.

They have managed to find partners who fund development and have a built-in potential-user base. In Chandler, it was Arizona State University, which offered space and underwriting of memberships for students. In Dearborn, it was Ford Motor Company, which also offered space and underwriting of memberships for its employees. In Paris, it was Leroy Merlin, a DIY home improvement chain, which offered space adjacent to one of its stores.

I attended the opening ceremony for the Paris location. The 215,000-square-foot shop, called TechShop Ateliers Leroy Merlin, is located in Ivry-sur-Seine. It has more than 150 machines over two floors, even more than in the flagship San Francisco TechShop. It's an impressive array of tools and workspaces.

"All these machines will be more accessible to more people," Stéphane Calmes said proudly, giving us a tour. He's the project lead for Leroy Merlin. He stood in front of a large-format UV LED printer made by Roland. Leroy Merlin customers might buy materials at their store, he pointed out, and customize them using the tools in TechShop. Stéphane showed an example of a door that was covered in a colorful custom print produced on the Roland printer. "This is something you couldn't do at home yourself." Another example was a small table bought at Leroy Merlin that had a print of the lunar landscape on its surface.

Stéphane walked us through the metal shop and then the woodshop, which featured traditional tools but also a ShopBot CNC cutter, locked inside a cage, as required by French law. He showed us the water jet, "the most spectacular and most expensive" machine they have. He explained that this TechShop has new safety innovations such as a set of sensors mounted on poles that can detect if anyone is standing near the water jet, and if so, not allow the machine to operate. There is even a bike shop.

Mark Hatch addressed the audience gathered for the opening. "The cost of tools has dropped ninety percent," he announced. Making them available for the cost of TechShop membership drops the end-user cost by an equivalent amount, he added. "TechShop is providing access to these tools of the new industrial revolution for the creative class. And now Leroy Merlin has brought the tools of this revolution to Paris." Leroy Merlin is working on two additional TechShops: one in Lille, where their corporate headquarters are, and Grenoble, in the French Alps. There are plans to build five TechShops in France.

TechShop has the potential to energize a maker community, attracting new makers and giving them the opportunity to experiment and innovate. Yet the for-profit model for makerspaces has been a challenge, not just for TechShop but for others like it who depend on members who pay to play.

FIRSTBUILD

FirstBuild is an open makerspace created by GE's Appliance division in Louisville, Kentucky. It is an unusual experiment: a corporate-funded makerspace open to the public but focusing on innovation in appliances. This makes FirstBuild the first vertical makerspace.

On a visit, I met with Kevin Nolan of GE, who developed the space, which opened in July 2014. The location is on the edge of the University of Louisville campus and several miles away from the GE's large Appliance Park. The facility has a public area that can be used for meetings or to display some of the work that goes on there. The work area is divided into two spaces: one for small-scale work like electronics, and the other area where the more dangerous equipment is housed. They also have a set of offices made from shipping containers. The entire space was built in six months. It is clean and well-organized, and yet not particularly corporate.

One of the GE team members, Tim, built a Raspberry Pi jukebox and speakers for the space, and Kevin commented that he had no idea how many different interests his employees had. The idea behind FirstBuild was to open up access to GE Appliances so that makers could innovate on top of them as a platform. Generally speaking, makers are not going to develop a new appliance and bring it to market. However, they might be able to modify existing appliances in useful ways. One example was adding a camera inside the refrigerator that could take a picture of what is in the fridge and send it to you, so you know what ingredients you have or what you need to buy.

Kevin explained the business case behind FirstBuild. He said that a company like GE, which recently sold the appliances division to a Chinese company called Haier, has to allocate a certain percentage of its profits to R&D—let's say four to six percent. The Appliance Division has an R&D budget, and its goal is to bring twelve products to market each year. He thinks FirstBuild can contribute four of those new projects a year and break even. The cost of the entire setup was not that expensive. There are eight people on staff, four of whom are skilled machinists that can help makers. Kevin believes he will be able to get new ideas and develop new projects that they would not have produced internally.

The FirstBuild space is open to the public for free, although you do have to become a member and sign a few forms. You can work on anything in the space, but it is clearly designed to encourage members to develop and test projects related to kitchen appliances—they have a professional kitchen there, too. FirstBuild worked with Local Motors to create an online platform for submitting ideas. Local Motors is the first car company to cocreate vehicles online with a virtual community of designers, fabricators, engineers, and enthusiasts from around the world, using open-source principles.

I liked that the initial submission was required to be tweet-like: short and sweet. As ideas get voted up, they can become a project, and GE can decide to fund the project, at which point GE will also help with manufacturing. Kevin said that his small team is having to reinvent small-scale manufacturing processes that had vanished from a large company. One of his motivations is to find new ways to keep the factory workers at Appliance Park employed. He said that they hope to develop new products like a start-up and be able to bring successful products to scale. One of the innovations to come out of FirstBuild is the Opal Nugget Ice Maker,[4] which raised $2.7 million on Indiegogo.

A big hurdle that Kevin overcame was getting GE lawyers to change their stance on the intellectual property that came out of FirstBuild. It was a pretty remarkable change. Members of FirstBuild retain the rights to what they develop with GE, and GE gets the right to use the invention if they pay a royalty. The maker can take the work elsewhere and even sell the rights if GE doesn't do anything with it.

Probably the first true collaboration between a company and a maker community, FirstBuild is also working with the University of Louisville's engineering students, who are on site in an adjacent area, as well as with LVL1, a makerspace in Louisville. LVL1 was started in 2009 by two teachers from Kentucky Country Day School who learned about makerspaces at a Maker Faire and decided to start one of their own. LVL1 is the opposite of FirstBuild—scruffy and disorderly, with little attention paid to what goes where or what is left behind. It looks like everything was donated, dragged in, and dropped where it is now. Yet they have about eighty members and are not worried about their ability to pay rent.

According to Ted Smith, chief innovation officer for the City of Louisville, when Kevin Nolan and his team started developing the idea for FirstBuild, they were going to locate it in either Connecticut or Arizona. Smith helped GE organize a hackathon to see if they could prove that there was enough talent in Louisville. At the event, which he said started slowly, several the LVL1 members began asking questions of the GE team, and something clicked. The GE team realized that they had talent in their own backyard.

Venkat Venkatakrishnan, one of the key people behind FirstBuild, told me about the thirty-six-hour mega-hackathon in April 2015. They had two hundred participants who organized themselves into thirty-two teams. Venkat said that forty percent of the people were from GE's Appliance division: employees who were showing up on their own time because they were interested. That's amazing.

The winner of the mega-hackathon was a project named House Roast. Two guys from GE had an idea about roasting coffee at home in a standard convection oven. One of them knew software while the other knew GE's wall ovens inside out. During the hackathon, they recruited two other team members based on their skill sets. All of them were driven by their own passion for coffee. Together the team hacked a GE convection oven to roast coffee, controlling the temperature with a custom roasting profile that was programmed into an Arduino. The team won the prize of $5,000.

The House Roast project proved the idea worked, but hacking an oven wasn't something everybody could do or wanted to do. So they launched an online challenge for a home coffee roasting kit that you could build and stick inside the oven and control the temperature with your smartphone. They got thirty-one entries.

The winner of the challenge was Stephane Arthur Kiss, a twenty-nine-year-old mechanical engineer from Ottawa, Canada, who was inspired by his popcorn maker to figure out a better way to roast coffee at home. Second place went to two engineering interns at GE, Hunter Stephenson and Steven Morse, ages twenty-one and twenty-two, who work at GE Appliances but worked on their designs at FirstBuild. Third place went to Chia-Chen Lee, thirty-two, an industrial designer living in

Rochester, New York, who came up with a magnetic levitating spinning drum to tumble the beans to roast them.

FirstBuild engineers will look at how to refine these designs and see if they can be manufactured as a product. The makers can choose to continue working on their project with GE or not. In this respect, the FirstBuild challenges function more like a design competition, and the award plus recognition is the goal of the challenge.

However, for GE, this process uncovered innovative ideas for a market niche that GE had never looked at. Venkat didn't realize that it was possible to roast coffee beans at home, and his colleagues hadn't realized how much passion there was for coffee-making. More engineers at FirstBuild began looking at solving other related problems, such as reducing the time it takes to make cold-brew coffee.

I can imagine that cities will have many different makerspaces that create a local ecosystem for makers, integrating hobbyists, businesses, and institutions such as universities.

UNIVERSITY MAKERSPACES

In addition to fab labs, there are a growing number of universities that have developed makerspaces that are accessible to students in departments such as mechanical engineering. Some are open to any student on campus, and some are even open to the community. In fact, universities have long had machine shops and art studios—the latter usually being closer to a makerspace in spirit than the former. But they had limited access, especially the machine shop, which was staffed by machinists often hired to do things for the faculty. Few of these spaces had the mandate to let students use the tools themselves.

At the Georgia Institute of Technology, the largest engineering school in the United States, the Invention Studio is a campus-wide makerspace open twenty-four hours to any faculty, student, or staff member and project, not just those in classes. The Invention Studio has $500,000 of equipment in three thousand square feet. One of the most innovative aspects of the space is the student-run Makers Club. Seventy students are members of the club, which provides support and training for the community of users. "We keep the space open and the machines running," said one student. It also gets students invested as stakeholders in the space and creates camaraderie. The Makers Club is currently sponsoring a grant program, offering $250 for student projects. What's especially positive about the program is that it is open to any student project in any major: "We love multidisciplinary projects."

Craig Forest, assistant professor of bioengineering, has been the faculty sponsor for the Invention Studio. He originally saw the makerspace as a high-end prototyping facility for students working on Capstone Design Expo projects, but it is becoming much more than that. There are now over five hundred users per month. Students don't just show up to work on class projects; they also work on their own personal projects. They also enjoy hanging out in the space. It has become their space, not just a space owned and operated by the school. When I visited, a janitor from campus was using the waterjet to create a metal frame for a drone he was building. It's a model that other universities should study and replicate.

At Southern Methodist University, the Innovation Gymnasium is in the basement of the engineering building. It is an open, flexible space, with twenty-four-hour access, and it is open to any student regardless of their department. The director at the time of my visit, Greg Needel, told me that he thinks of it as a "third place" on campus, apart from the dorm room and the classroom. It is also the other "nine-to-five" space, meaning 9 p.m. to 5 a.m., because students are typically there very late.

Greg has created a picture wall, taking Polaroid photos of the two-hundred-some students who use the space. He wanted to make it easier for students to find others who used the space based on their interests and their projects. This relates to what I see as a greater need in the community, the need for maker portfolios. There is a need to know what other people are doing, and what kind of work is in progress. This fosters collaboration and interdisciplinary thinking.

At Case Western Reserve University in Cleveland, Think[box], a workshop for collaboration and innovation, is a seven-story, fifty-thousand-square-foot makerspace available to students and faculty but also open to the public for free. It received $10 million in funding from alumnus and adjunct professor Larry Sears and his wife, Sally Zlotnick Sears. Larry founded an electronics company called Hexagram that automated remote meter-reading for utility companies.

Ian Charnas has been the driving force behind Think[box], which got started in a four-thousand-square-foot basement space. He is a graduate of Case and explained that he just hung around wanting to do this until they made it his job. He gave me a tour of the new facility before it was even entirely built. While Think[box] is affiliated with the Case School of Engineering, Ian explained that Think[box] is not just an engineering space but will have a broad focus that includes art and fashion, and they hope to work closely with nearby institutions such as the Cleveland Institute of Art.

One project I saw was Felipe Gomez del Campo's plasma-assisted fuel nozzle, which "improves the way fuel burns in jet engines" and was featured at the White House Maker Faire. Other projects included a foot-powered cell-phone charger, a 3-D magnetic skull puzzle, and a rapid malaria detector kit.

Ian explained that the seven floors have been designed to accommodate different stages of development for a person or a project. The first floor is dedicated to community, a gathering place; the second floor is for ideation, with lots of whiteboards and open space for brainstorming. The third floor is for prototyping, while the fourth floor is for fabrication—don't ask me to explain the difference. The fifth floor is an open projects space, essentially workbenches and storage. The sixth floor organizes resources for entrepreneurs, and the seventh floor serves as an incubator for small groups that form to develop a new product.

One of the students, Kailey Shara, is a super-smart young engineer who left Case to cofound Carbon Origins, one of my favorite maker start-ups. Carbon Origins is building an Arduino-compatible flight controller called Apollo, built to guide a homemade rocket and survive the inevitable crashes of test launches. She has returned to Case to complete her undergraduate education, and she was clearly delighted to see the progress with the new Think[box]. I could almost see her thinking about how much time she would be spending there.

Ian Charnas mentioned a survey in which 39.9 percent of the students indicated that Think[box] was a significant factor in their decision to attend Case Western Reserve University. I believe we will find this to be true across more colleges and universities. It will provide a clear rationale for investing in student-accessible makerspaces.

Add to that the benefits of developing an innovation ecosystem that engages the community and partners such as manufacturers, investors, and corporate research organizations. It has the potential to broaden the base of support as well as the social and economic impact of these makerspaces.

A MAKERSPACE IN SHENZHEN

Chaihuo Maker Space was the first makerspace in Shenzhen, sponsored by Eric Pan of Seeed Studio. It is a small two-room space located in a building that had been a factory and is now a collection of retail shops and galleries in an arts-and-design district known as Oct-Loft. With lime-green walls and bookshelves, Chaihuo seems more like a drop-in social space than an industrial workshop.

In January 2015, Premier Li Keqiang, the number-two official in the Chinese government and known for leading economic reform, visited Shenzhen and stopped by Chaihuo Maker Space to meet with Eric and other makers. He must have liked what he saw because his visit set off a chain reaction. An article in *China Daily* said that "Li called on Chaihuo to show the nation's growing commitment to supporting grassroots innovation and the budding Maker Movement that is producing it."

I was told by more than one person in Shenzhen that in China, a single sentence from a powerful person can be all it takes to set things in motion. An article titled "Maker Culture to Be Encouraged in China" cited Li Keqiang's visit and said that in March the State Council established initiatives "for fostering creativity and entrepreneurship." The initiatives encourage investment in small and medium-size businesses. Li Ouya of the Chaihuo Maker Space was quoted in the article: "Maker culture used to be like underground water, a subculture thing here in

China, with the water seeping up from the ground. But since the premier's visit, it has grown into more like a well, with more and more different maker spaces surfacing and taking root across China."[5] At the Shenzhen Forum in April, Deputy Mayor Tang Jie of Shenzhen talked about makers fostering creativity and innovation in Shenzhen:

> Currently, Shenzhen is the most active city in venture capital and private equity investment in China, so the city can support, embrace, and understand makers. Makers also should learn to share, and they need a restless, youthful spirit, to cultivate their abilities through practice, and keep creative and imaginative. Shenzhen is to be built as the city of makers.[6]

The local Shenzhen government organized support for Shenzhen Maker Faire in June 2015, and every school in Shenzhen will receive $50,000 to develop a makerspace. The challenge, even with such government support, is still how to build a community around these makerspaces.

WORKING OUT AND SCALING UP

What's next for makerspaces? Whether for-profit or nonprofit, future growth depends on having more paid staff, a trend toward the professionalization of makerspaces. The 2014 Makers Nation survey reports that twenty-nine percent of the one hundred makerspaces in the survey rely only on volunteers, and fifty-four percent have one to five paid staff.[7] For a makerspace to grow and serve a broad community, they must be able to perform a core set of services to support membership expansion. New people need to be welcomed when they arrive. Classes and workshops are essential to provide basic safety training as well as offer workshops for members who arrive without project ideas.

Good governance is also key. This amounts to building an organization, not just creating a space. "Good governance is resilient, consistently serves its constituency, and is predictable," said Travis

Good. It's something that Travis and I talk a lot about, as there were few basic questions that we found ourselves asking to understand how a space works:

- Is there a board of directors, and does the leadership change over time?
- How are decisions made about growth, such as moving to a larger space or adding equipment?
- How invested are members as stakeholders in the space, such that they support it and have a voice in how it is managed?

The development of makerspaces can be compared to the rise of fitness clubs. Today's health clubs started out years ago as bodybuilding gyms. They were designed to meet the needs of a narrow, largely male membership. They weren't particularly friendly to newcomers or casual users. Yet something changed in our culture around physical fitness, which became acknowledged as a key aspect of health and well-being, and gyms had to become more open and accommodating if they were to grow. They had to learn to welcome women as well as men who did not see themselves as bodybuilders, attracting both serious and casual members. This is the kind of change we can expect as makerspaces grow.

For all the emphasis on equipment, what makes a makerspace are the people who run it and the community of people that it serves. There is a difference between a place that people use and a place where people feel they belong. Getting that culture right is important—creating an open, supportive culture for exploration, risk-taking, creativity, and personal development. A good makerspace can be transformative for its community.

5

How: Components, Tools, and Markets

How people make things is changing for a number of reasons—mostly because the tools of production have become more accessible and affordable. This includes cheap electronic components; standardized microprocessor boards that connect and control things such as sensors and LEDs; and digital fabrication tools that combine software with hardware, creating the object on a computer and then sending the instructions for building the object to a machine such as a 3-D printer or a laser cutter. These tools and technologies—some of which are described in this chapter—are driving the Maker Movement.

Tools that were once available only in industrial settings, R&D labs, or university departments can now be found in many of the community makerspaces I described in the previous chapter. Small desktop-size digital fabrication tools can replace the large equipment used in factories, and what was unimaginable not long ago—making one of a thing—is now feasible. "The tools of making have never been cheaper, easier, or more powerful," Mark Hatch of TechShop likes to say. This democratization of technology means that more people can afford to get access to tools, learn to use them, and surprise us by making things that would never have come out of the industrial world.

The word *technology* comes to us from the Greek root *tekhnē*, meaning "craft," "skill," "art," or "technique." Fire is usually considered the first technology that humans used: a tool itself for heating and cooking, which was then used for making other tools. We can consider cooking a technology too: a set of tools and techniques that transform raw materials into cooked food. The same is true of sewing. The Singer sewing machine was the first "domestic" machine. Technology involves learning new skills but also imagining new applications. The power of a new technology resides not just in what it does but how it makes us think. As Marshall McLuhan said, "We shape our tools and our tools shape us."[1]

MATERIALS AND COMPONENTS

Makers find parts and materials from a variety of sources. Take duct tape: I'm frequently amazed by the variety and ingenuity of the creations, such as wallets, hats, and dresses in a duct-tape fashion show, not to mention the bridges, boats, and hammocks that people have built with the stuff. When school administrators claim they don't have the funds for a makerspace, I like to tell them that origami—using recycled newspaper, even   io a great activity for young makers.

I first heard the word *obtainium* when touring the San Francisco workshop of Mark Pauline, the founder and director of Survival Research Labs. Mark and a team of contributors build large machines that come to life as part of ritual performances. When I asked where he gets the parts, Mark told me with a wink that they use "obtainium," loosely defined as abandoned, recycled, unused pieces of equipment. An artist might call them found materials.

Many of the makers I know are scavengers. They visit salvage yards, flea markets, and garage sales and spend time on Craigslist and eBay looking for materials. It's part of the appeal of making to hunt down the right part or find something that you didn't expect to find and wonder what you might do with it. Marque Cornblatt, the robot maker and drone racer, started with assemblages of stuff he found on the streets of New York City or dumpster diving. When he moved to the Bay Area, he found a different set of things in the trash, especially around Silicon Valley, where the material he found was more high-tech. That was when Marque started making more sophisticated mechanical and kinetic pieces.

In Louisville, Kentucky, Mayor Greg Fischer wants his city to be a leader in reducing, recycling, and reusing materials that otherwise end up in landfills. The city's electronics recycling center is a place where you drop off old computers, cell phones, and printers. It's a place where makers might go if they have ideas about how to use such equipment and its components. For instance, old printers are a good source for servomotors, useful for robotics projects. The town dump can hold unexpected treasures. Many makers take pride in being resourceful, finding things that other people no longer use or value.

At Maker Faire Tokyo, a Japanese maker asked me a question that was troubling him: "How do you make things in America? You don't have Akihabara." In Akihabara, there are large and small stores selling cheap electronics and new and used gadgets, but what makes Akihabara special is that you can also find all the components that are used inside those gadgets. You have to navigate a warren of cramped stalls, each one specializing in a single component, such as capacitors, inductors, or switches, loose and sorted in boxes by color, size, and type. Each stall is like an educational catalog of what exists that you might use. It's a fun, if jarring, place to visit, which I have done several times in conjunction with Maker Faire Tokyo. Each time I've seen something I didn't know existed.

Then there's Huaqiangbei, the electronics market in Shenzhen, China, reputedly the largest in the world. It's featured on the Shenzhen Map for Makers. In a city known for its factories and local supply chain, the SEG (Shenzhen Electronics Shopping) Market in Huaqiangbei is its heart. I got a tour from Bunnie Huang, an American living in Singapore who is a frequent visitor to Huaqiangbei. Bunnie is a hacker-hero, a MIT PhD who wrote *Hacking the Xbox,* developed the Chumby as a hacker-friendly consumer device, and works on projects like Safecast and the open-source laptop. To follow Bunnie through the many levels of multiple stores in Huaqiangbei is like being led by an expert hunter on a trail through the jungle. He knows where things are, who the people are, how to get in, and how to get out, all while avoiding the crowds. What is in focus for him is a blur to me: copycat cell phones, power supplies, test equipment. Bunnie comes here to source parts for many clients, but he is also looking for parts for his own projects. Here is the SEG Market in Bunnie's own words:

As I first step foot into the building, I am assaulted by a whirlwind of electronic components. Tapes and reels of resistors and capacitors, ICs of every type, inductors, relays, pogo pin test points, voltmeters, trays of memories, all crammed into tiny six- by three-foot booths with a storekeeper poking away at a laptop, sometimes playing go, sometimes counting parts. Some booths are true mom-and-pop shops, with mothers tending to babies and kids playing in the aisles.

And it's not like, oh, you can get ten of these LEDs or a couple of these relays like you do in Akihabara. No, no. These booths specialize, and if you see something you like, you can usually buy several tubes, trays or reels of it: you can go into production the next day. Over there, a woman sorting stacks of 1 GB mini-SD cards like poker chips; here, a man putting sticks of 1 GB Kingston memory into retail packages; next to him, a girl counting resistors.[2]

Bunnie and I stopped at one booth where a woman in a green dress was actively soldering on a workbench. She stopped her work to talk to us, as Bunnie opened a bag of LEDs and tested several of them before eventually asking about the price. I left with several bulk packs of color LEDs, LED lighting strips, plus a cheap but powerful cell-phone charger. Each transaction was negotiated by Bunnie. He can get the best price and find the broadest selection in this market, and he knows that these transactions only happen in person.

In the United States, there's no real equivalent to Akihabara or Huaqiangbei, certainly no district that I know of in any city. Most people think of Radio Shack as a place to get cheap electronic components, and yet they moved away from that business to sell phones and remote-controlled toys. They stopped being that kind of place until the Maker Movement made them rethink what they were doing. For the holiday season in 2014, Radio Shack ran a wacky ad campaign featuring Weird Al Yankovic singing and standing in front of a store wall with Make-branded kits. Unfortunately, it was too late in the game, and unable to reduce its debts, Radio Shack declared bankruptcy in 2015.

The answer I gave to the Japanese maker who asked me about Akihabara was that American makers buy components online from websites like Digi-Key and McMaster-Carr. He shook his head, as if to say that it's not the same as Akihabara. Online sites for electronic parts, and even their phone-book-thick print catalogs, can be as intimidating as visiting the electronic markets in Asia. They were developed with a professional audience in mind, one trained in sourcing parts. You have to know exactly what you want, and the number of choices can be overwhelming for a first-time maker venturing out to look for LEDs or sensors.

A group of maker-friendly and beginner-friendly online stores such as SparkFun and Adafruit have provided a valuable service, somewhat like Bunnie himself, by knowing which components are most likely needed for maker projects. If you are looking for an accelerometer, Digi-Key will have hundreds of them, while SparkFun might offer a half-dozen and provide documentation on how you might use them in a project. Adafruit goes a little farther, designing and developing more of their own components, which reflect the engineering acumen of founder Limor Fried.

The availability of components influences how makers think about what they can build. Yet being able to find parts is one thing; finding them available for cheap is another. That's why some clever makers also look at new consumer electronics devices, because they introduce new technology into the market at a relatively low price. A collection of new devices, including the Wii Remote, the GoPro camera, and the Microsoft Kinect, have caught the attention of makers who have different ideas than the average consumer about how to use them.

Sometimes, as in the case of Microsoft Kinect, the manufacturer takes issue with those who open up the device or hack them to get access to new sensors and other parts inside. The Microsoft Kinect, which came out for the Xbox game platform, is a remarkable set of cameras and sensors that can detect motion and allow interaction with a game console without using a keyboard or hand-held game control. Makers immediately could see other uses for the Kinect that were never conceived of by the game designers at Microsoft.

When Microsoft released the Kinect in the fall of 2010, Limor Fried of Adafruit announced a $3,000 reward for the first person to create an open-source driver for the Kinect. The software interface would allow applications not tied to Microsoft to access Kinect and control it. Hector Martin, a Spanish hacker also known as Marcan, bought a Kinect in the store, and three hours later had hacked it, posting a minute-long video demonstrating the feat: a computer running Linux interacting with the Kinect. Martin won the bounty offered by Adafruit, and his code eventually became part of an open-source project called OpenKinect. A website was soon launched called Kinecthacks.net.

Microsoft reacted strongly to the invitation to hack Kinect and even denied that the hack happened. Then, as they began seeing the wide variety of applications appearing because of the hack, Microsoft changed course and said that they had designed the Kinect to be hacked. Makers discover uses that manufacturers haven't designed for, which the manufacturer can see as a legal issue, or the start of an innovation ecosystem.

Hacking the Kinect unleashed all kinds of applications that have been created over the years since. I mentioned Phoenix Perry's Nightmare Kitty, which used the Kinect along with a machine learning library, to challenge kids to move and avoid the falling kitties. One favorite was a medical application that allowed a patient at home to monitor their own range of motion by raising their arms up and down in front of the Kinect. Willow Garage, a Google-backed robotics company, rewired its telepresence robot to use a cheap Kinect as its vision system in place of a very expensive custom camera system. A number of makers have used it to do full-body scans of people, which they can send to a 3-D printer to create personalized figurines.

The Kinect is an example of a device that becomes a component, and moreover, a component around which a community develops. It becomes a standard, which is to say, a well-known way to do a wide variety of projects. Makers benefit from using standards over custom solutions because they ultimately save time (and headaches). You don't have to figure out things yourself, and even if there are problems—and there usually are—you can rely on the fact that others are encountering those problems and also trying solve them.

LADYADA, CHAMPION OF OPEN-SOURCE HARDWARE

Inspired by the open-source movement in software, makers began to apply open source to hardware, eventually creating the open-source hardware license. While an open-source software license describes what to do with source code so that anyone can copy or modify the program, an open-source hardware license describes a hardware design that allows anyone else to replicate the design. It might include schematics, a bill-of-materials for all the parts, software source code and other design files, and documentation. The intent as described in the Open Source Hardware Definition is that the hardware design "is made publicly available so that anyone can study, modify, distribute, make, and sell the design or hardware based on that design."[3] Open-source hardware grew out of using the Internet to share the plans for a project so that others could build it if they wanted.

Limor Fried began to learn about electronics as well as open-source software at MIT while she was getting her bachelor's and master's degrees. Like many MIT grads I've met, she comes across as completely self-taught, which is all the more credit to MIT. With her distinctively punk pink hair and bottom-lip piercing, Limor is a pioneering developer of open-source hardware products. She has a remarkable presence, especially when giving a public talk. She's her own deeply geeky person and her charisma can really surprise you. Listening to her, one feels she can do absolutely anything she sets her mind to—and she's proven that she can.

With personal projects that she created, Limor began sharing schematics and other files online, often going by "LadyAda," an homage to Ada Lovelace, who wrote the first program to be executed by a machine. One of her first projects was called Spoke POV, a custom circuit board with LEDs that created an image using a persistence-of-vision device placed on a spinning bike wheel. Limor had not thought much about licensing or patents or copyrights. She put up her files under a noncommercial license, meaning that others could copy and use the designs for any reason except to create a commercial product. "It doesn't matter if you put up something with a noncommercial license," said Limor, "because someone probably will come along and make a derivative [work] anyway, and it will be closed source, and he might be your ex-boyfriend." That actually happened to her. The noncommercial license "really backfired" for her.

Limor's next project, which she did to learn microcontrollers, was the Minty MP3 player built inside an Altoids mint tin. "Building MP3 players was the hot new thing" in 2003–2004. She could find lots of examples of these DIY MP3 players online, and it inspired her to want to make her own. "Because people had started to put up files, and said that they didn't know what license it would be under, they concluded the files were free for you to use in any way you wanted." She relied on these designs to build her own—a derivative work.

"Because this project was based on the generous donations by all these people of code and examples," she decided she should give back. She made her own files free for others to use. And they did, including those who took the designs and manufactured a product for sale. "I was a little upset at first, because I didn't even consider that possibility," Limor said. "But it turns out that it worked out anyway, because then Apple came out with the iPod."

At that point she had learned enough about electronics that she decided to build a cell-phone jammer for her thesis project. A cell-phone jammer sends out a signal that disables a cell-phone's reception. Because it is illegal to manufacture, sell, or use a cell-phone jammer, Limor considered the project simply an exercise to demonstrate that one could be built. She figured she had arrived at a design that no one else would rip off.

Years later, however, I was on an Amtrak train from New York to Boston with a maker who is an associate of Limor. After listening for too long to a loud person on a cell phone, the maker pulled out a homemade cell-phone jammer, hidden inside a cigarette box. The maker pushed a button and instantly ended the cell-phone conversation several rows ahead of us. I could see the person with the phone look quizzically at it, unaware of what had happened. I immediately had a vision of an arms race escalating between cell-phone users and cell-phone jammer users. Being party to the prank was a guilty pleasure.

After graduating from MIT, Limor started her own open-source business in 2006. "It doesn't seem to make any sense to do open-source hardware," she said. "If you do something good, it will probably be stolen from you. If you do something bad, you'll be made fun of." Yet she decided from the moment she founded Adafruit Industries that she would share her design files and code for each product she created. She didn't think it would hurt her sales, and part of her didn't care if it did. She believed in doing it and accepted the risks. That conviction, forged from her own experience, is what makes her such a pioneer in the Maker Movement.

Limor was named Entrepreneur of the Year in 2012 by *Entrepreneur* magazine, and she was the first woman engineer on the cover of *Wired*, although missing in her cover photo were her pink hair and lip ring. They had her in the familiar pose of Rosie the Riveter instead.

Limor now designs, engineers, manufactures, and documents her own products. She created a factory in Manhattan with pick-and-place machines, and she manufactures the Arduino board in the United States, in addition to her own product lines. She has built a thriving company, Adafruit Industries, that tripled and doubled to reach $40 million in sales in 2015, using an open-source hardware model and no venture capital or other outside funding. What worked for Limor was not just designing great products, but her insistence on openness, sharing, and community.

Open-source hardware is more than a license. It is an understanding, a kind of informal handshake among members of a community. It is an agreement that can have a legal meaning but also implies an unwritten code about how to use the work of others and share back your own work. Phil Torrone, who left his editorial position at *Make:* to work with Limor and become her partner, wrote about the "unspoken" rules of open-source hardware: "What does the open-source maker want?" he asks. "Just to be credited properly." He waggles a finger at those who use "open" as a marketing pitch but don't actually share their designs. If you say your product is open-source, don't require an NDA to get the source code or create other hoops for people to jump through. Also, if you use open-source code and designs, add value to them. Don't just give them a different home with a new name. Finally, he warns those who just want to clone to "go do something else."

These rules for participating in the open-source hardware community are not arbitrary. They exist as ways of demonstrating good faith in the community, valuing its work, recognizing the reciprocal benefit of learning from one another's expertise, and contributing back something of one's own. It's about playing well with others, whether for fun or profit.

Another way of looking at it is how a group of musicians might learn songs from each other and play them in addition to each person creating their own music. That can be done freely, requiring some kind of acknowledgment that you are playing someone else's music, even as you adapt it. However, if you were to record that music for purposes of selling it on an album, you should not only credit the source but find a way to agree on a royalty for using that music. In music, it's required, and there are organizations that enable and enforce the payment of royalties. The Open Source Hardware Association doesn't have the resources to enable standardized transactions for royalty payments or the ability to enforce them. Strictly speaking, the license doesn't require it. Yet understanding the difference between personal use and commercial use is important to maintain the open-source hardware ecosystem.

Applying open source to physical products is an important breakthrough, not just for electronics. Marcin Jakubowski's Open Source Ecology project in Missouri promises an open-source tractor and other industrial equipment. There are open-source clocks and watches. Opendesk provides open-source furniture designs. WikiHouse is an open-source platform for building designs. OSVehicle is developing an open-source hardware platform for electric cars.

The spread of open-source hardware is related to the increasing use of software to generate physical goods, as well as code embedded in those products. Open-source software and hardware combined ensure that we have the freedom to understand how things are made, replicate them if we choose, or just modify them if we wish.

ARDUINO

A typical maker project consists of a bunch of parts and a brain to control them. That brain is a small microcontroller, a computer on a single electronic board the size of a credit card, with pins that can be wired to the outside world for input or output. It's often referred to as a "board." Microcontrollers in and of themselves are not new; specialized ones are embedded in all kinds of devices, like cars, implantable medical devices, televisions, and kitchen appliances. Each microcontroller might have its own unique design and programming. Microcontrollers have been widely used in industrial applications, most of which are proprietary. Now, a growing number of microcontrollers including Arduino, Raspberry Pi, BeagleBone, and Propeller represent the next wave of a personal computing revolution. Used by hobbyists who are starting small, this new hardware comes from unexpected places, designed by people with a real sense of purpose.

Arduino is Italian. It happened to be the name of a bar frequented by art students from the Interaction Design Institute Ivrea near Milan. In 2005, an associate professor at the Institute, Massimo Banzi, offered the name of that bar for a simple new microcontroller aimed at artists and other nontechnical users. With the introduction of Arduino, makers had a cheap, open-source microcontroller that was easy enough for nontechnical people to use. Arduino became a standard "brain" for millions of projects, and a large community grew up around it.

In March 2012, I spent several days with Massimo, first at a Maker's Conference in Rome and then in and around Turin, where the Arduino is made. Massimo is not an engineer, and that's what makes the story of Arduino so fascinating. Most microcontrollers are designed by engineers for engineers. Not Arduino. It was specifically designed for designers taking courses in interaction design, which he explained to me:

> Interaction design is the design of any interactive experience. It can be the interface of an object, say a device with three buttons. That interface can yield an unsatisfying experience, or those same three buttons can create a really positive experience. Everything

is an experience, an interaction between you and something, and that experience can be designed.… In my case, interaction design tends to be about technology. A lot of the experiences that you have today are enabled by objects that contain electronics and sensors. Technology enables the communication between you and the device, or you and a service. The interaction designer must be a designer but must also understand technology enough to know what kind of experience you can create with a certain tool.

Yet Massimo was struggling with teaching electronics to his design students. "When I explained what electrons were, for instance, Ohm's law, they didn't understand. Then I realized that that's not how I had learned. The way I learned was experimenting. When something didn't work, I would go back and try to understand why it worked. So that theory became useful to me, and it matched reality. Then I started to teach like that and make everything much more hands-on."

What Massimo wanted with Arduino was a cheap and uncomplicated way to control sensors, lights, sounds, motors, and other elements of what could be a museum exhibit, a performance, or an article of clothing.

The students usually don't have a background in technology. They don't know how to program or to do electronics, and we only gave them two to four weeks to create physical comput- ing projects. At that time, the tools you'd find in the market were mostly designed for engineers, with a lot of options, lots of jumpers, and lots of connectors. Students found them too complex and couldn't figure them out properly.… So the idea was to make a board with the minimum number of parts with a cheap price tag. I wanted them to cost $20 for a board: the price of a pizza dinner. So a student could afford to skip a pizza dinner and spend the money on a board.

A master's student advised by Massimo, Hernando Barragán devel- oped the Wiring project for his thesis, and his work led to Arduino. Others helped with designing the board and the interactive development environment (IDE) for programming the board. The core Arduino team consisted of five people: Massimo; Tom Igoe, who wrote the book on physical computing and taught at NYU's ITP program, which has goals similar to Ivrea; David Cuartielles, a Spaniard, who was teaching interactive technologies at Malmö University in Sweden and first got involved to design the Motor Shield; David Mellis, at MIT's Media Lab, whose focus was developing the IDE; and Gianluca Martino, an Italian, who was responsible for manufacturing the board in Turin.

Massimo explained, "The whole idea of being a maker involves concepts of collaboration, community, and working with other people. It's very hard to be a maker and be by yourself locked in a room or even in a lab. It's really something that involves a lot of collaborations at different levels. Arduino boards are a mash-up of open technologies wrapped up in a unified user experience."

On my visit, we started one day at the fab lab in Turin, then we hopped in Massimo's Fiat and drove into the countryside, which reminded me of where I live in Sebastopol, California, except that it is surrounded by the Alps. This Piedmont area is where the early personal computer maker, Olivetti, once thrived. The factories that make Arduino, and the people behind them, are industrial descendants of Olivetti. They set up shop on their own after Olivetti closed its doors. This is why the region has the machinery for making printed circuit boards.

Seeing the manufacturing process first-hand, I was struck by how much the process that creates a printed circuit board is like a printing press. The board starts out as a copper-clad laminate sheet that goes through a variety of different processes, involving CNC machines for drilling holes, photochemicals, silk screening, and hot ovens. At the end, there's a stack of blank circuit boards in the characteristic Arduino blue, the signature look of Arduino, in which Massimo takes great pride. The boards go to another factory where pick-and-place machines choose components that are fed on reels and then are placed onto the board by a robotic hand. The boards then require additional tasks, some of which involve soldering by hand. Finally, all the boards go through extensive testing. The finished boards are then packaged in boxes and sent to a warehouse, where they are eventually shipped out in bulk.

The Italian owners of the factories take great pride in what they do and the fact that they do it in Italy, seemingly against all odds. They can boast that not only do they make the boards in Italy, but the machines that make the boards are also Italian.

The first run produced two hundred preassembled boards, fifty of which were bought by Interaction Design Institute Ivrea. Another fifty went to Sweden, and the remaining hundred sold out quickly. "From then on, we had people asking us for boards. When I started to see what people were doing, I knew that Arduino was making a difference," said Massimo.

He believes that the Arduino reflects a minimalistic design philosophy:

> Good design is about using the minimal amount of stuff that you need. If something is visually simple, it encourages people to use it. If you look at the stuff that engineers design, they tend to be large printed circuit boards with lots of buttons, switches, and lights on the board. If I'm a beginner and I look at this thing, I see thirty buttons, jumpers, switches, different configurable items. Our brain in a completely subconscious way starts to multiply all the options together. It might conclude that this thing in front of my face has three billion combinations, which is way too many to comprehend.
>
> This why the Arduino is incredibly streamlined, including the very minimum number of parts that you can use. We removed every jumper because each one has the potential to confuse people who are trying to learn. That's a design choice. Certain people believe in complex devices with lots of features. Others believe in simple things.

From these origins, the enthusiastic interdisciplinary mix of engineers thinking like artists and artists thinking like engineers, working together and learning from each other, Arduino has become a tinkering platform for all kinds of people. It began with modest ambitions. It's not the most powerful microcontroller. Its virtues are being cheap—around $35—easy to use, and open, meaning you can freely share hardware designs and code, and you can use it with any OS. Each of these virtues is important, but being cheap is first.

Cheap means you can try out Arduino with little investment. You don't have to know in advance whether it will do what you want, or even know exactly what you want. You can experiment and find out without risking a lot of money. An Arduino board is cheap enough that you wouldn't feel bad breaking it, burning it up, or leaving it behind embedded in a project. You wouldn't do that with an iPhone or a PC, but you can do it with Arduino.

Arduino shares its design files for the board layout as well as its parts list so that anyone else could replicate it under a Creative Commons share-alike license. It allows others, even companies, to create their own boards using the same design or to modify it. However, the Arduino team trademarked the name Arduino. This means that anyone can clone the Arduino, but they can't call it Arduino. Massimo distinguishes between clones and counterfeits, the latter being boards that appear to be Arduino boards using the name and logo but are not actually authorized by the team.

Massimo's five-member team agreed to jointly own the trademark, which was the basis for a business growing the Arduino ecosystem. In 2015, Gianluca Martino broke from the group, claiming he owned trademark rights to Arduino in Italy and the United States, and sold those rights to investors. It has turned into a legal battle that is still not resolved, and unfortunately it has caused confusion in the community about what Arduino means. Massimo has begun releasing boards under the Genuino name in countries like the United States where the ownership of the Arduino trademark is in dispute.

PERSONAL FABRICATION

There are three kinds of personal fabrication machines that are almost always found in fab labs and makerspaces: laser cutters, CNC cutters, and 3-D printers. These digitally driven machines provide new ways to make things but also encourage new people to become makers. Manufacturing gets personal when you can do it on the desktop, not just in a factory.

GLOWFORGE

I've heard makers say that 3-D printers get people to come to a maker-space, but it's the laser cutter that keeps them coming back. A laser cutter is a 2-D cutting machine that uses a laser to cut through materials, typically wood but also cardboard and textiles. The laser actually burns through the material, just as you can use a magnifying glass to direct the rays of the sun to burn a piece of paper. The focused laser beam is digitally controlled to follow a path for cutting or engraving. Anything that you can draw as a 2-D image or pattern on a computer can be sent to the laser cutter. Relative to 3-D printers and CNC machines, laser cutters are much easier to learn and use productively—that's why people come back. The only real drawback to laser cutters has been their price.

Dan Shapiro thought he could build a laser cutter that was considerably cheaper but just as capable as more expensive machines. Dan is as high-energy as the lasers in his machine, and he lights up instantly when talking about Glowforge. "This is a device I have been craving for years," he said. "It has all kinds of superpowers—the brain lives in the cloud; it has a bunch of on-board sensors, and a CO_2 laser—all in a desktop form factor." The target price for Glowforge is about $4,000, while other laser cutters range from four to ten times that price. When Dan offered a presale price of fifty percent off, he raised over $20 million, demonstrating huge demand for this kind of machine.

A former Microsoft employee still living in Seattle, Dan sees the desktop laser cutter, or 3-D laser printer, as he likes to call it, as a creative tool with many more uses than 3-D printers. He sees crafters using it for leather or textiles. It's possible to draw on fabric and have the machine detect the pattern and follow it for engraving or cutting. This is because Glowforge does a 2-D and 3-D scan of the material and can recognize a drawing. The same design could be applied to paper, wood, or cardboard.

There have been several different attempts to develop an open-source laser cutter. One was called Lasersaur, a large-scale laser cutter from Addie Wagenknecht and Stefan Hechenberger of Nortd Labs in New York City. Another was the Risha project, a portable laser cutter designed by Moushira Elamrawy from Alexandria, Egypt. Dan was able to garner more resources to design and build his machine. "The open-source projects use off-the-shelf components such as lasers and a power supply," he said while at Glowforge, they have designed many of their own components. Dan has been able to hire engineers with a lot of experience to help keep the price point low but also design for safety. "Glowforge is as safe as a DVD player," said Dan, who adds that the U.S. Food and Drug Administration has to approve any device with a laser, and "no laser light can escape the device." It's almost like a kitchen appliance. When you open the top of the laser cutter, the laser stops immediately. It's perfect for school and library makerspaces.

Carl Bass, CEO of Autodesk, said he is really impressed by Glowforge. "We need better machine tools, machines that are smarter," he said, adding that machines need to have a feedback loop and be able to "close the loop." He means that machines should be able to recognize a problem when it happens and begin trying to solve it for you. For instance, a 3-D printer doesn't know that a print job is failing, and it keeps on printing. A person must intervene and stop the print job. Carl thinks that today's machine tools require that the user be an expert to solve problems, but users shouldn't have to be experts. "We need machine learning for machines," he said.

SHOPBOT

When I was looking through online applications for the first Maker Faire in 2006, one that caught my eye was from an Oakland couple who said they had a CNC machine in their dining room. CNC stands for computer numerically controlled. In short, it's a type of cutting machine that is controlled by a set of computer-generated instructions rather than manually. Jeffrey McGrew and Jillian Northrup, an architect and a graphic designer, were using a ShopBot CNC machine to run a design-and-build custom furniture studio that they called Because We Can. Once they were accepted as exhibitors at Maker Faire, they wrote ShopBot asking if the company would be willing to send a CNC machine to the event. Even though ShopBot couldn't have known much about Maker Faire, they said yes, and Jeffrey and Jillian's booth had a full-size CNC machine in operation during our very first Maker Faire.

In 2007 I went to visit Ted Hall and his team at ShopBot in Durham, North Carolina. Ted was a professor of neuroscience at Duke University who said that his "hobby has always been wanting to build boats—backyard, plywood boats." He paused thoughtfully and rubbed his reddish beard. "Not that I have finished any yet," he added, with a self-deprecating laugh.

Ted set out to make a wooden boat but got so sidetracked by building the tools needed to make the boat that he never accomplished the original goal. Ted has an unfinished boat sitting in the rafters of his garage, yet he has a fifty-person company that sells several different models of CNC machine: ShopBot. The company grew out of his efforts to build a machine that could interpret instructions from a computer design model and guide a robotic tool's cutting path. As he explained, it is not just that a CNC—sometimes called a CNC router or a CNC cutter—automates what a woodworker might do manually; it allows you to make shapes that would be very difficult to create otherwise.

> It occurred to me that you could make much more attractive plywood boats using CAD [computer-aided design] software than anyone had been able to do in the past. My starting point for getting into plywood boat building was to create some software. So I created an add-in for a DOS CAD program. What it was able to do was to allow you to sculpt with plywood and not have it end up looking like a shoebox. With plywood, you would bend it and then cut it where it ended on something square, but the curve of it really wants to take a different shape.

I didn't quite understand what he was saying until I looked more closely at boats and how they are built. The sides of a boat have to curve in order to come together as a joint at the prow. He asked me if I knew the term *developable surfaces*. I did not. He explained,

Developable surfaces are the form that sheet material such as plywood takes when you bend it. Most 3-D modeling programs work on the idea that you can infinitely manipulate 3-D surfaces. But you can't form a sheet of paper into a globe. You've got to slice it and bend it. *Those* are developable surfaces. You need software to model those surfaces because it's so difficult to do with traditional drafting techniques. Then, once you have represented the curved surface, you need to make it flat again, which will give you the lines to cut in the actual plywood. If you glue them together along those lines, then the object takes the 3-D form that you designed it for. It's a lot of geometry.

Ted started to build his first boat, an eight-foot rowboat. He was printing out sheets from a printer, taping them to the wood, taking a saw, and cutting the outlines. "Why am I cutting this by hand?" he thought.

I went around North Carolina looking for a used, beat-up CNC machine. Back then, they cost $30,000 to $40,000 for one that I could never make work anyway. Now, what I should have done is bought one, and I would have saved myself a whole lot of misery. But instead I figured: *I can make one of these.* CNC machines were driven by stepper motors, and I had a lot of experience running stepper motors in my science career. One of the things we did a lot in the lab was use a stepper motor to drive an infusion pump: basically using a motor to cause a pump to inject something into a rat's brain or its stomach.

It turned out, however, that the motor wasn't the hard part. "What's hard is having a mechanical system that's rigid enough to do what you want to do and still be affordable." A CNC requires precision to do its job by moving one location to another. Vibrations might cause the material or the cutting tool to move out of alignment. It is a harder problem than Ted imagined. After ten years working on it, he was only now feeling that ShopBot's mechanical system "was dialed in."

His first version was a kit. It had instructions, drawings, and a parts list. Customers would have to build the physical machine themselves, and Ted would send them the electronics for the machine. The parts were mostly available at Home Depot. "However, one of the limiting factors was that the machine required thirty patio-door wheels, and usually you can get only about twelve of those at one store," he said. The demand wasn't immediate, though: "I figured that if people could have their own CNC machine for a couple hundred bucks, then everyone would want one. It turns out that wasn't true." It was still too hard to build one of the machines by yourself, even with a kit.

Ted gradually began supplying more of the parts, but it still required a lot of assembly. "I always saw these machines as for the backyard do-it-yourself guy or for small shops." He began advertising them through tiny ads in woodworking magazines. "We had a CNC machine that was one-tenth the cost of other machines, so professional woodworking shops were our early adopters. For five years we didn't sell one that was used more for play, for DIY. So this market said that they'd be happy if it was a quarter of the price of the big guys, but they wanted it put it together, and they needed it to be faster and easier to use." Ted now sees ShopBot as a low-end disrupter of much larger industrial companies who sell CNC machines.

Ted talks about the challenge of selling a computer-based woodworking tool to woodworkers who have little experience with computers. Often they had a sense of losing control, rather than gaining control. Makers tend to be the opposite: they understand computers, but they don't know woodworking as well. ShopBots can now be found in almost every Fab Lab, TechShop, and many larger makerspaces.

Hall's unfinished boat is still waiting for him in the barn. Meanwhile, customers like Because We Can are using ShopBots to build custom office furniture installations, such as the one with a Captain Nemo theme that they designed and built for a game development company in San Francisco. Jeffrey and Jillian eventually moved the CNC out of their living space and into a warehouse in Oakland.

3-D PRINTERS

It seems like magic: in 3-D printing, a digital file leads to the creation of a physical object. The essence of 3-D printing is that a computer can "slice" a 3-D image into a stack of 2-D layers, and a machine builds the object by adding one layer on another; thus 3-D printers have come to be called "additive" manufacturing, which forced CNC to be considered, awkwardly, "subtractive" manufacturing. It can be mesmerizing to watch a 3-D printer "robot" in action, but it is also tediously slow, taking anywhere from one to six hours to build objects any larger than the palm of your hand.

The first patent for 3-D printing dates to 1980s, around the same time that the laser printer was patented by Xerox PARC. Charles Hull was the inventor of a prototyping process to make three-dimensional objects out of plastic. In an interview at Maker Faire in 2013, Charles told me about the problem he was trying to solve back in those days: "It took six weeks to several months once you had a design to get a plastic part, and then typically it was wrong and you had to do it over." He was working for a company in Southern California, and he told the owner that he thought he could build a machine to solve the problem.

"I was enthused about it, but the owner wasn't," he said. "He finally agreed that if I would do the work on my spare time, he'd give me a lab in the back. Which is what I did. I spent lots of evenings and weekends in this lab, making a lot of stuff that didn't work." Eventually, he had a working prototype of a three-axis machine that directed an ultraviolet light to harden a plastic coating, dot by dot, building up an object layer by layer.

Charles also came up with STL file format, which were the set of instructions that operated the machine. In his prototype, Charles originally wrote instructions line by line to tell the machine how to make the object, which was tedious. He understood that CAD programs could be designed to generate the STL file for a design, which could be exported for the 3-D printer.

He got the patent in 1986 and founded 3-D Systems that year. The first customers for 3-D printing were large companies: automotive, health care, and aerospace. "The early technology wasn't very good." However, the early beta customers provided valuable feedback and funded development so that "we quickly developed really solid equipment."

The early printers were large, expensive, and required technical expertise to operate. So 3-D printing was confined to large manufacturers and few industrial design shops that had applications that justified the cost of the machine. There was no consumer market for 3-D printing, although Charles said it had been a hope of his. Industrial 3-D printers might be compared to mainframe computers: it took nearly twenty years for the analogous "personal computer" for 3-D printing to emerge by radically reducing its size, complexity, and cost.

MAKERBOT

The first efforts to build a 3-D printer for the rest of us was an open-source project called RepRap, short for replicating rapid prototype. The project was founded in 2005 by Adrian Bowyer, a senior lecturer in mechanical engineering at the University of Bath in the United Kingdom. Its goal was to design self-replicating machines, meaning machines that could build themselves. By 2007 the first version of a RepRap machine, called Darwin, was released and slowly gave birth to a hobbyist community of RepRap builders. As an open-source project, RepRap builders shared ideas and designs freely. RepRap gets credit for creating the foundation on which most consumer 3-D printers were built. The trouble with RepRap was that building a working machine was just plain hard to do, yet it did not deter the enthusiasts.

The promise of an open-source 3-D printer that was affordable and easy to build captivated the makers who tried to build RepRap models. Most didn't really know why they wanted a 3-D printer and what they would actually do with it. They were willing to be pioneers and do the hard work so that they could have access to one and then figure out what it could do for them. That kind of spirit led to the founding of Maker-Bot by Zach "Hoeken" Smith, Adam Mayer, and Bre Pettis, who had met at NYC Resistor, a Brooklyn-based hackerspace. Zach had built a

RepRap model and believed strongly in open source while also seeing an opportunity to create a commercial 3-D printer that was accessible to more people. He teamed up with Bre, a former middle school teacher of art who joined *Make:* magazine to create "Weekend Project" videos and developed a strong following. It seemed like a perfect recipe for success. Bre saw himself as Steve Jobs, the promotor, while Zach was the Steve Wozniak, the technical wizard behind the scenes. By 2010, Bre was on the cover of *Make:* volume 21, holding MakerBot's first product, the CupCake 3-D printer, which was sold as a kit. MakerBot took off, and it both created and stood to capitalize on the emerging 3-D printer market; 3-D printing became red hot.

PRINTRBOT

In the race to bring 3-D printing to the masses, MakerBot was the hare and Printrbot was the tortoise, and there were lots of tortoises. Brook Drumm and Printrbot could be viewed as an also-ran or a winner in the long-run. Time will tell, but he's still going along at his own pace.

Brook grew up the son of a pastor, living on a farm in northwest Ohio near an Amish community. "I am made to take things apart, understand how they work, and put them back together," said Brook. "I was made by God to be a maker." He recalls that the house he grew up in was centered around the kitchen table, which was handmade by his father. The family dinners were "raucous."

Brook, who is bald and wears the long goatee of a biblical character, has pretty strong convictions. He intended to follow in his father's footsteps, so he went to a Bible college in Kansas and became a minister. He took a job in the Sacramento area to work for a mega-church where a classmate had also landed. Brook was in charge of the multimedia systems in the church: multimedia shows for Sunday services were a pretty big attraction. Then, unexpectedly, the church laid him off and he wasn't sure what he would do.

As he was thinking about what he would do next, he came across the issue of *Make:* with Bre Pettis of MakerBot on the cover. As he looked at it, he thought Bre had just shown him that anyone could do this, that he could build a 3-D printer himself. "I saved for six months and bought a MakerBot Cupcake Kit for $799," said Brook. He had to advance himself $200 from his credit card, which he did without telling his wife, Margie. He built the kit on his kitchen table. He taught his son, who was six, to solder.

Perhaps the most important thing Brook realized was that there were others out there just like him. He organized the first 3-D printer meetup in the small town of Roseville, California, calling it the NorCal RepRap and MakerBot Builders Group. He brought his eldest daughter along, partially so that he'd have company if no one showed up. He also wanted her there so that she could see "the start of something from nothing," if indeed he was successful at starting something. Brook recalled that it was just the two of them for a while, and he wasn't sure anyone was going to come. Then two people showed up and he felt relief, especially because both of them were compelling. One was a crafty old-time programmer who had worked in a language called Forth that Drumm had never heard of. The other was a transgendered woman named Stephanie who worked at Intel and knew about designing electronics.

At subsequent meetings the group grew to about sixty. "If I could attract sixty people in Roseville who are interested in 3-D printing," said Brook, "then what about the rest of the world?" He had a fourteen-year-old boy show up who was also building a Cupcake 3-D printer. A woman who called herself a prostheticist came to a meeting because her job was to make "eyes, ears, and noses" and she thought 3-D printers might be useful for that. "At one meeting, I asked how many people wanted to build a 3-D printer for $800 and nobody raised their hand," Brook said. "So I lowered the price to $600 and still nobody raised their hand." He then asked how many people wanted a 3-D printer, and everybody raised their hand. "That's when I knew that price was really important." He began to see his opportunity.

Brook decided for a future meeting that he would order parts for ten RepRap 3-D printers and just get people building them. He put that on his credit card, again without telling his wife. He had gotten some work doing Web development, but he didn't have money to spend freely. He bartered a trade with an auto-body shop in which he did their website and they allowed him to use a conference room once a month for his meetings, which would now involve the actual building of 3-D printers. He neglected his family, working on designs in the garage for eight months. He taught his wife how to do coding for the Web so that she could take on his development work and he could focus on 3-D printers.

"My wife didn't want to be talking 3-D printers all the time." She told him to move the 3-D printer, which was now running from the top of the washing machine, to the garage. She had grown tired of hearing it running all the time.

In what he calls his "aha" moment, he was sitting out in the backyard, trying to get some air and take a break from this new obsession. It was 2 a.m. He had an idea for a new design for a 3-D printer that would simplify it and possibly reduce the cost in half. The new design would open up the machine because it was no longer enclosed in a box. The bed moves back and forth on bars; it was no longer stationary.

The next day, Brook went to a Lowe's Home Improvement store. "I didn't have money to spend," he recalls. "I was looking in the reject bin and I found a four-by-four post, a piece of wood you'd use for a fence post. It cost me $2." He brought it home and drilled two holes for a Y-axis and two more for a Z-axis. It was enough for him to know that his new design would be stable. He ordered additional parts, and soon had a prototype. He brought it to the 3-D printer group in Roseville. "When they first looked at it, they said it wouldn't work, but then they tried it, and it did work." He convinced six people in the group to help him build ten more of his design. Stephanie, the Intel engineer, looked at his electronics and said that she could do something a lot better, and did.

Given that he had a working prototype, Brook decided to raise money on Kickstarter, launching a campaign in November 2011. In the campaign he promised "to simply do my best to make the most incredible little 3-D printer ever." He said it's a 3-D printer that a kid could put together. His goal was to raise $50,000 to build and deliver fifty machines. In eleven hours, he met that goal. His campaign exceeded his expectations, raising $831,000 in one month. He was thrilled that people trusted him.

However, it was not all good news. First, Kickstarter took its cut and so he got $750,000 in his bank account. Next he learned that because he was not incorporated, the amount he raised would be taxed as income. He would have to write a check for $333,000 in several months to the IRS. "I will always remember that number because it was such a shocker," he said. He also now had to deliver over 1,100 Printrbots, far exceeding the number fifty that he had felt comfortable that he could do.

Brook wrote to his backers that he now had to move production out of his garage and into a rented warehouse. It was in the brick-walled basement of an old train stop, beneath an Old Towne Pizza. The first thing he did in the space was build a table with a few people he recruited from the group to work with him. The table would be the central feature of the office, just as his father's kitchen table was the center of his home.

As he worked on producing the first batch of machines, he created videos to show their progress to his Kickstarter backers, but this didn't satisfy many people, who were expecting immediate delivery of a machine. Malicious rumors were spread that Brook had skipped town with the money and now was in Mexico. "I thought I was thick-skinned, but the whole experience made me thin-skinned," he said. Margie bore the brunt of it, trying to help answer questions. "It crushed her," said Brook. "People can be so mean." Printrbot began shipping units to its Kickstarter backers in August 2012. "Nine months. Not too bad compared to some other Kickstarters."

Many people he had met through the meetup group came to work for him, as well as two sons of the pastor who had let him go from the Sacramento church. One of the group's best and brightest, Karl, an HP scientist and engineer, was making the "hot ends" or extruders for Printrbot. "He's probably made more hot ends than anyone, over sixty thousand." Brook has never visited the place where Karl makes the hot ends, but he knows that Karl has a workshop on his farm where he raises bees and llamas.

Brook opened a Printrbot store and began taking orders. He sold $200,000 worth of printers in the first month. Then PayPal decided that Printrbot must be defrauding customers, and they seized his funds. Brook needed the money to build the machines and pay his employees.

For months, he argued with PayPal. One person there became his nemesis. "I put up a web cam just to show him what we were doing." It took months, but eventually he got the money.

Brook stayed in the basement space for a year before moving to a warehouse near the municipal airport in Lincoln, California. It is flat and barren country, hot in the summer. On a Friday afternoon when I visited in 2013, there were six people working there, including Brook. It is a rough space. There is no air-conditioning for summer, no heating for winter. Yet it is a full-fledged manufacturing facility with three large laser cutters doing most of the work. They have one person running those jobs, another one assembling and testing, and a third doing shipping. I met Caleb, a seventeen-year-old who does a lot of the development and programming. Brook and Caleb share a small office where it is quiet.

Brook has not raised any venture capital. He has funded his own development out of cash flow. He remains a leader at offering a reliable product with reasonable performance at a low price, about one-quarter the price of a MakerBot. Once he visited a Sand Hill venture capitalist to whom I introduced him, just to explore the possibility. In a large conference room, the VC began the meeting: "Tell me, Mr. Drumm, what is your superpower?" Put off by the question, Drumm didn't really have an answer. At the end of their meeting, the VC said: "I know what your superpowers are, Brook. First, you are good at identifying talent and getting good people. Second, you are obsessed with product design." Brook got something out of the meeting, and it wasn't financing, but he was happy to get back to his humble warehouse.

In 2014, MakerBot was acquired for $604 million.[4] I asked Brook what he thought about MakerBot's acquisition. He said he found it both scary and motivating. He sees it as an opportunity for him. Brook has fully embraced open source and the community in the same ways that MakerBot once did. He said he's learning from Bre and MakerBot about what to do and not to do. Brook met Bre once, at Maker Faire New York. Bre said to him, "Hey, I saw your printer." And then added: "One thing, though, you aren't charging enough for it." Brook said he was really nervous, but wished he had said in return, "I think you're charging too much."

He's had the lowest-price product on the market and feels that his product quality was high. Over four years, he has grown to $8.5 million in revenue. While a small operation, he and his team are doing quite a lot, and they're very innovative. Mostly recently, Brook released a new kind of CNC machine that he calls Crawlbot.

Brook is representative of a set of small-scale makers of 3-D printers, like Diego Porqueras of Bukobot in Pasadena, California, or Rick Pollack of MakerGear outside Cleveland. Like them, he is at a crucial point in time when he must make key decisions about how to grow and compete. Brook has conviction that he should stay focused on what he can do.

Most objects created on a personal 3-D printer are made of plastic. At the industrial level, however, there are a wide range of materials available, including metals. An artist could perfect a prototype sculpture using plastic in her own studio, then send out the model to a service such as Shapeways that could print it in solid gold.

A prototyping revolution that is making rapid development and iteration possible for all kinds of products is being led by 3-D printing. Physical prototypes can be tested in the physical world and handed over to other people to inspect. Rapid iteration means that you can improve a design before committing to its manufacture. Not only is prototyping cheaper, but also more people are able to afford to do it. We are just beginning to find out what these new capabilities mean, for applications that range from toys to medicine, musical instruments to houses.

THE MAKING ECOSYSTEM

It's not just the tools that are revolutionizing the process of making. Today, makers have access to online platforms like Instructables, which I described in chapter 3, as well as Thingiverse and YouTube, all of which offer blueprints, designs, or step-by-step instructions for how to do and make all kinds of things. On top of that, makers everywhere have access to a set of resources for finding investments and funding, on the one hand, and for selling their finished product on the other.

In many cities, particularly around the holidays, craft fairs and maker's markets have sprung up, taking their cue from farmers' markets. A farmers' market serves a local community of food producers and a broader community of people whose interest in fresh local food helps to support those producers. The producer can have a direct relationship with customers, who value the personal interaction. The lesson of farmers' markets is that they have become an informal gathering place for the community.

In Milwaukee, there is a Maker Market on the first Sunday of the month in a parking lot next to a coffee shop. Its website says: "These Sundays have come to support a rotating cast of talented artists, crafters, makers, and designers from Milwaukee and beyond, selling everything you can imagine two hands being able to create—from soy candles and small batch, hand-mixed beauty products to one-of-a-kind clothing and jewelry."

Then there's Etsy, which was launched the same year as *Make:* magazine as an online platform for selling handmade goods. It has all the features of a craft fair, except that anybody from anywhere can visit. A buyer can feel that the transaction is taking place directly with the producer, and that items that are in limited supply are seen as more valuable.

Rob Kalin and a team of two or three people designed and built the site. I met Rob in 2003 while the site was in development. He saw an opportunity to create a market for handmade goods that you would never find at Walmart or Amazon. Investor Fred Wilson of Union Square Ventures (USV) said that Rob once told him, "I'm an artist. Making websites is my medium right now."[5] Wilson and USV were among the original investors in Etsy. Rob was intense and idealistic about Etsy and its community. He was the right person to get it started. He was first CEO, then chief creative officer, then left the company, came back as CEO, but was eventually fired.

Executive coach Pam Klainer tells the story of Rob giving a talk on entrepreneurship at a New York business school:

> He was introduced by an exceptionally dry and dull professor who read an exceptionally dry and dull definition of entrepreneurship.

Rob took the microphone, waved his hands in the air, and said something like, "No, no, it isn't that. It's having a passion for an idea and making it into something you could never have imagined when you started."[6]

For buyers, Etsy is an endless row of boutique shops where one-of-a-kind craft and creativity shine; for sellers, it is an easy-to-use platform to list inventory, take sales, calculate taxes, and streamline other tasks related to entrepreneurship—with a built-in audience of millions. Through Etsy, makers can reach a market that might not be available locally.

Two-thirds of Etsy sellers are women who use their shop to supplement their income. Many are starting a business on the side, and it is work they can do from home. Making one's first sale can be a magical moment. An Etsy store can provide powerful feedback and encouragement. The lessons of entrepreneurship and running a business can be learned with a rather low level of investment. For some, the success they have on Etsy can lead to opportunities elsewhere, which might involve setting up physical storefronts or licensing products to manufacturers.

Etsy's blog editor Michelle Traub said about the community:

> The key to success on Etsy is telling your story. Nineteen million shoppers from around the world come to this marketplace, inspired by a fundamental desire for objects with meaning. Our community has turned away from anonymity to have a conversation at the farmers market, discover the history of a vintage find, or buy directly from a maker.[7]

For years, Etsy required that its sellers produce their goods by hand, or sell used second-hand items they curated. This policy fit the production methods used by most crafters, but makers who used a variety of machines to make things couldn't use Etsy to sell. Also, once a seller became popular on Etsy, it could be difficult for her to keep up with demand making everything by hand. In 2013, Etsy changed their policy to allow sellers to produce goods made by machines in a factory. Located in Brooklyn, New York, Etsy went public in 2015 with a valuation of $1.8

billion. At that time, they had fifty million members and its platform facilitated $1.93 billion in transactions.

In many ways, Kickstarter, which is known as the top crowd-funding site, is also a marketplace, although its language shifts from sellers and buyers to creators and backers. On Kickstarter, makers can share their product idea and seek backers to fund it before actually having built the product. Kickstarter requires them to show a working prototype, as some way of ensuring that the person's idea is real. Kickstarter is, in effect, a form of pre-selling, which means taking orders before you have the product. Kickstarter has resisted this idea, and its founders wrote a blog post in 2012 titled "Kickstarter Is Not a Store." Other crowd-funding sites like Indiegogo and Crowd Supply have proven to be viable alternatives as well.

Kickstarter works surprisingly well for makers, allowing them to test out whether there is a market for a new product. It's an effective way to launch a new product before it is actually available for sale. Of the makers who exhibited at the 2014 Maker Faire Bay Area, 123 of them had used Kickstarter and collectively raised over $23 million. That includes makers like Lisa Qiu Fetterman, creator of the Nomiku sous-vide cooker, who were able to use Kickstarter not just to get funds for the development of a product but also help to create a market for it.

Taken together, all of these innovations create a vibrant ecosystem in which the way things are made has fundamentally changed. With access to this ecosystem, anyone can take an idea and prototype it, get feedback on it online from a global community of experts, then tweak it and prototype it again, over the space of a week or a month.

6

Toy Makers

If you want to prepare a generation of children for a changing world, change the toys they play with.

> Hello! To the Boys and Girls of America. The Gilbert Hall of Science is yours from this hour on. For me it is a dream come true—for you I hope it will be an inspiration, a guiding line in the middle of that seldom smooth road which leads the young and struggling engineer, chemist, railroad builder, scientist, research worker ... to studying and learning, ever striving, that success at the end of the road is the only true success. Every student and builder who works for the sake of humanity finds in the end that he has worked mightily for himself, too.

Alfred Carlton Gilbert published this welcome message in 1941 at the opening of the Gilbert Hall of Science, a multistory showcase of toys located in New York City at Twenty-Fifth Street and Fifth Avenue. Gilbert was born in Salem, Oregon, in 1884 and had gone east to get a medical degree at Yale that he never used. He was interested in three things, he said: "athletics, sleight-of-hand, and scientific experiments." He won the pole vault in the 1908 Olympics, tying for the gold medal, after inventing a better pole as well as the pole vault box that caught the pole and secured it for the pole vaulter to rise in the air. (Before that, the pole had a spike at the end of it that dug into the ground.) Gilbert's first company was Mysto Magic, which created a series of magic kits, a business that was barely profitable. Then, while making frequent train trips from New Haven, Connecticut, to New York City, he was inspired by the steel girder construction of skyscrapers and bridges to create a new kind of educational toy, the Erector Set. He began producing the first sets in 1913, and they became an immediate success.

The Erector Set was one of the central toys of my childhood. It took its place alongside the other great American toys of the age: Tinker Toys and Lincoln Logs. All were developed around the second decade of the twentieth century, and became very significant construction-based learning sets. Between 1913 and 1966, thirty million Erector Sets were sold in the United States. The Erector Set's decline followed Gilbert's death in 1961 and the bankruptcy of the A. C. Gilbert Company in 1967.

The popularity of the Erector Set spanned the technological era from the Ford Model T and the electrification of America to the age of aerospace, and the Erector Set evolved to keep pace with these developments. It reflected the optimistic, can-do spirit of the American Century, of a society that was rapidly gaining new abilities to solve problems and do ambitious projects because of science and technology. The Erector Set was an invitation for children to participate in that future—and do work with their own hands.

The Erector Set box was filled with steel girders, a small battery-powered motor, and the parts to make wheels and pulleys. Erector Sets were available in versions numbered 0 to 8. The higher the number, the more pieces there were. These kits contained the parts to build specific models such as a train bridge or a Ferris wheel. In the 1920s, the number 8 Erector Set cost $70 and weighed a staggering 150 pounds; it included all the parts to build a five-foot zeppelin.

The Erector Set was an ideal toy for the ideal American boy, whom Gilbert defined as competitive, clever, and curious, just like him. Perhaps the first to create advertising that appealed directly to boys, Gilbert spoke to them as a friend and mentor, characteristically opening his ads with the greeting "Hello Boys." His slogans for the Erector Set included "Young Boy's Paradise," "1,000 Toys in 1," and "World's Greatest Toy."

Today, Gilbert's American boy seems a little homogenous and corny, like a Norman Rockwell painting, but we still recognize him in ourselves and in our children—the girls too.

Historian Bruce Watson's biography of A. C. Gilbert, *The Man Who Changed How Boys and Toys Were Made,* claims that Gilbert was not merely the inventor of the Erector Set; he transformed the image of the American boy from problem child to problem solver, from delinquent to constructive contributor. The Gilbert Hall of Science was not only a place but an idea: that if you give children the right tools, or toys, they will educate themselves.

TOY STORIES

We never quite let go of the toys we played with as kids. I enjoy asking makers about the toys of their childhood. Makers remember talking bears, bubble blowers, toy boats, racers, robots, and View-Master 3-D slides, all of which inspired them to see the world differently, as something they could shape, mold, shrink, and hack.

"I loved Shrinky Dinks as a kid," exclaimed Michelle Khine, a biomedical engineering professor I met at a weekend summer camp for scientists called SciFoo. Michelle explained how Shrinky Dinks inspired her to come up with a new nanoscale process that led to her start-up, Shrink Nanotechnologies. By creating a design at larger scale and then shrinking it down, Michelle was able to find a simple and inexpensive method of making microfluidic channels for what she calls a "lab on a chip." Her early prototypes were printed out on a laser printer and then baked in a toaster oven. One use of this process was to create saliva-based assays for infectious diseases.

Arduino cofounder Massimo Banzi, introduced in chapter 5, learned electronics through the Braun Lectron kit, a German-made electronics kit that used magnetic blocks to build circuits and connect other components. Many people fondly remember the Radio Shack 150-in-One Electronic Project Kit, which was like a large game board that used springs as connectors. Craig Smith of South Milwaukee, Wisconsin, writes on the *Make:* blog: "I was one of the lucky kids that received a Radio Shack Science Fair 150-in-One Electronic Project Kit on Christmas morning. I spent hours making the different projects, such as sound effects, radio, and light experiments." He said that he is in the process of recreating this beloved kit in wood as a work of art.

I once enjoyed a fancy dinner prepared by The Cooking Lab's Nathan Myhrvold and Maxime Bilet, authors of the epic cookbook, *Modernist Cuisine at Home.* The book is an impressive effort to understand and explain the science of cooking and how to use such knowledge to develop new techniques and recipes. The Cooking Lab team works in a kitchen inside a machine shop inside an R&D lab in Seattle. For this dinner, there were more cooks busily preparing the meal than people eating it. One of the last courses was Nathan's take on Gummi Worms, which were made from a gel infused with olive oil, vanilla, and thyme. The gel was poured into a mold used for making commercial fishing lures. Eating them was a delight, turning us into kids dangling wiggly worms above our mouths. It reminded me of food-making toys like Incredible Edibles, a 1960s-era Mattel toy whose secret ingredient was called Gobble Degoop. I remember dozens of molds for making insects with frightening appendages, all of which you could eat.

One maker who looks at toys for his parts is José Gómez-Márquez, director of the Innovations in International Health Lab at MIT. He has been rethinking the design and deployment of medical devices in developing countries. According to José, most of the medical equipment comes to these countries secondhand from developed nations. Often, the conditions in the field require that practitioners customize or hack these devices to make them work. In addition, it can be difficult to find replacement parts and make repairs.

"When you need a part, you don't have access to McMaster-Carr or any parts supplier," he said. "Yet there's an amazing supply chain for toys, so you can find them everywhere." Going to a toy store might be the best option to find a part. "From a toy helicopter, I can find a rack and pinion system." José takes a DIY approach to medical equipment for developing countries, designing kits with parts for creating devices that can easily be adapted to field conditions.

At the Exploratorium's Tinkering Studio in San Francisco, Karen Wilkinson and Mike Petrich have organized workshops on toy dissection that invite children and adults to examine what's going on inside their store-bought toys. They set out a collection of toys along with a variety of hand tools such as scissors and handsaws. Then, without much direction, participants are encouraged to rip, cut and snap open these toys. Talking plush animals and dolls, as well as plastic noisemakers like a Speak & Spell, can reveal a circuit board, speakers, and even sensors. Once you have identified and isolated these components, you can put them together in new ways, creating a new toy from components harvested from old toys. If you want to try this at home and don't want to take apart your favorite toys, you can find toys in bins at Goodwill and other thrift stores.

One of the more unusual toy hacking exercises is called circuit-bending, which involves opening up any battery-powered toy that makes sound and short-circuiting its electronics. In *Make:* volume 4, Christiana Yambo and Sabastian Boaz wrote about how to "bend" a Casio SK-5 sampling keyboard and turn it into a different kind of musical instrument.[1] In the same article, they introduce the father of circuit-bending—the person who coined the term—Reed Ghazala. In 1967, while Ghazala was in junior high, he was craving his own synthesizer but couldn't afford one. In his desk drawer he found a "junked mini amplifier made by Radio Shack, with the battery installed and the back off, exposing the circuitry."[2] After he closed the drawer, he heard strange sounds coming from inside. He realized that a metal piece had touched the circuit board and it was shorting out. His lightbulb moment was: *What if you shorted circuitry intentionally?* That's how circuit-bending was born: by accident.

Most people would consider the sounds generated by circuit-bending to be "distorted" electronic music or, less politely, noise. As we have organized accomplished circuit-benders at Maker Faire, I've learned that circuit-bending is less appreciated as music and more appreciated as a way of making strange sounds, in the same family as the Theremin. Said another way, as much as circuit benders enjoy how they can make this music from children's toys, few people want to listen to it, although it can be fun for a while to see a circuit-bender playing music using the knobs, sliders, and controls of a plastic toy. What circuit-bending demonstrates on another level is that we can modify and hack the internals of a machine and change its interface to suit our own purposes, none of which the manufacturer intended us to do.

CARDBOARD CREATIONS

I have a two-year-old grandson who, like many kids, enjoys playing in cardboard boxes, pulling the flaps in on him, as though it were a vehicle. Cardboard is one of the simplest, most widely available building materials, and its uses for young makers are incredible. All a pile of cardboard requires, to paraphrase Thomas Edison, is a good imagination.

As a nine-year-old, Caine Monroy hung out for summer vacation at his father's East Los Angeles auto-body shop, where he began building out of cardboard what he called Caine's Arcade, which was open on weekends to the public. The arcade was a set of imaginative games, all made from cardboard boxes that were leftover in the shop. Anyone could play the games for a little bit of cash and win prizes. Most of the prizes were Caine's old toys.

Filmmaker Nirvan Mullick came upon the arcade while visiting the shop to buy a part for his car. After talking to Caine, he bought a "fun pass" for $2, which allowed him to play five hundred games. Nirvan decided to make a video of the young maker, whose sweet, playful nature comes across well on film. The video went viral, and Caine's Arcade became a sensation, sharing a message of the importance of creativity in children's lives. The popularity of the video led to an annual Global Cardboard Challenge that invites children to participate by creating new things out of cardboard and other recycled items.

Juniper Tangpuz is a grown-up artist (in his thirties) who creates intricate, kinetic sculptures out of paper and cardboard. I met him at the Kansas City Maker Faire, where he was smartly dressed, wearing a bow tie and suspenders. He showed me an expressive 3-D puppet head with animated eyes and mouth. He peeled open the head to show me that the structure was made out of a gift box and a lunch bag. Nearby was a black-and-white cardboard bear with needles sticking out of its skin, which he called the Cactus Bear. Another was a hard-to-describe musical instrument called Piano Face, in which a cardboard collar holds piano keys so that when they are pressed, several soft-foam hammers hit your cheek to make sounds with your mouth as a resonator. He demonstrated a cardboard violin that made no sound, but when bowed, activated a spinning zoetrope that had images of a dancer in motion. Born and raised in Kansas City in a family of Filipino descent, Juniper told me that he was one of those kids who had to keep busy, and his parents gave him paper and scissors so he could "make things from stuff lying around the house." His art is now featured in galleries and museums.

Caine Munroy and Juniper Tangpuz show that simple materials can be perfect to allow one's imagination to take over—something that some of the most expensive toys fail to do. Makers' special appreciation for toys is leading to some of them to become toy makers themselves. Makers are bringing a new sensibility to play, sometimes inspired by their own experiences as children and other times from what the technology allows in the design and development of a new product.

3-D PRINTED CREATIONS

Wayne Losey was a toy designer at a large toy company who left because he found that it was too hard to create new toys within the company. He said that Star Wars was the best thing and the worst thing for the toy industry. It was good because almost any toy that licensed the Star Wars name and characters was very successful. It was bad because the mainstream toy industry saw licensing as the secret to success, and subsequently innovated less with new toys. A company that sells mass-market toys wants to make sure it can sell millions of a toy from its initial release. Gone were the days, according to Wayne, of creating a new toy that might take a while to catch on and build a following. It was a better business to dress up older toys and make them look new. A good example is the Star Wars version of Monopoly.

Wayne left because he felt that the "toys were dumb," and the ways the toys were made was dumb. He set out to create his own toy company. "I wanted to put toys in kids' hands that surprise and challenge them," he told me. "I wanted to build tools for creativity and imagination." His goal became finding ways for kids to "become their own toymaker."

On his own, Wayne set out to learn new skills such as 3-D modeling software and unlearn some of what he knew from working at a large company. "I had to shake the idea that I was a professional in the toy industry."

His first product was called ModiBot, a modular snap-together system for creating 3-D-printed action figures. Wayne came up with a design for interconnecting elements and a base set of parts as a starter kit. New parts could be designed by the community of users and shared for download on home 3-D printers. Eventually, he teamed up with several developers to create a new company called Modio that produced an iPad app that made it easier to design the characters. He launched the product at Maker Faire Bay Area in 2014.

Wayne realized that he was way ahead of the market: too few people have 3-D printers at home. Even he was using a 3-D printing service, Shapeways, to create his base set of parts. He believed that this kind of customization was where the toy industry would eventually go, and he wanted to help lead it there. It was opening up new ways to think about designing and developing toys, and putting children at the center of the process. Modio was acquired by Autodesk in 2015 and renamed Tinkerplay, becoming more of a software product for designing your own playthings.

Alice Taylor had a similar notion that came to her while she was walking around New York Toy Fair in 2010. The fair is where the toy industry shows off its new product lines to retailers. While checking out the small space in the basement featuring digital toys, the idea struck Alice that the future of toys would combine physical and digital interactions. She wondered if she could create a system where kids could design their own doll character on a computer and then send the design to a place where a high-end 3-D printer could make it for them. She founded Makie Labs in East London and raised some money to build a software team and set up manufacturing. I visited Makie Labs early in 2015 to see six or seven people managing jobs at 3-D printers, in a small factory that not only produced the doll figure but also its hair and clothing, and then boxed it and shipped it anywhere in the world.

The key for Alice's product is easy-to-use design software that gives children the ability to create an unlimited number of variations of a doll, changing shapes and colors to make a doll uniquely their own. Like other maker products, success depends on setting up a collaborative community for sharing. In that community, users might share clothing patterns for their own doll's wardrobe. The dolls are also designed to incorporate wearable electronics. While more expensive than an American Girl doll or mass-manufactured dolls like Barbie, Alice Taylor's dolls are truly one-of-a-kind, and they might serve a niche market of geeks and their progeny. Alice believes that the costs will come down, and that the appeal is broad, even bending the notion of what dolls can be for boys as well as girls.

Alice knows that the large toy companies are paying attention: she meets their representatives at Maker Faire who come "because it is a fertile ground for new emerging toys." Makers like Alice Taylor are doing the advance work of innovation for the toy industry of proving that such a product can be made and that kids want it.

CREATIVE CIRCUITRY

The most fundamental element of electronics is creating a circuit. A basic circuit consists of a battery, some wire, and an LED. Batteries have a positive and a negative output; LEDs have a long and short wire. They have to be put together using a simple logic test that it works. While you can use standard electronics components to build a circuit, there are other, more creative ways, too. Instead of messing with wires and breadboards, you can build soft circuits using conductive tape, thread, or paint.

One of the more fun examples is Squishy Circuits, developed by AnnMarie Thomas and her student Sam Johnson at the Playful Learning

Lab at the University of St. Thomas in Minnesota. Squishy Circuits uses homemade playdough to replace traditional wires. It requires making two batches of your own dough, which involves heating a mixture of water, flour, oil, and cream of tartar. In one batch, salt is added as an extra element to make a conductive dough. In the other, sugar is added to make it an insulating dough. Children already know how to use the dough, so they can sculpt all kinds of shapes, but they can also learn how to attach a battery and LEDs to build a circuit that powers the LEDs. I have seen fourth-graders in a public school in the Bronx, New York, figure out things for themselves, adding buzzers and motors to the circuit while shaping it into creatures or sculptures. Squishy Circuits are a lot more tactile than working with wires, especially for little fingers, and the results are more expressive. Instead of a jumble of wires and components, the kids talked about something they made.

I once had several batches of dough in my carry-on at the airport, and it set off the alarms at the TSA checkpoint. The dough was wrapped in plastic, and one batch had been dyed pink. A man from the bomb-defusing unit arrived to look at my bag and interrogate me. "What is it you do?" he asked me. "I do science experiments," was my answer, which seemed to satisfy him enough to let me go. My advice is to leave your playdough at home.

Paper craft can also be combined with conductive tape to create pop-up greeting cards that beep or light up. Marie Bjerede, a tech executive in Portland, leads this activity at her children's school as well as at the Maker Faire in Portland. She uses construction paper, scissors, glue sticks, coin-cell batteries, LEDs, and a roll of copper tape. The point of her activity is that a circuit is only a mechanism that you can use in service of what you want to express in the card. A pair of red LEDs might act as the eyes of a dragon or a robot.

Jie Qi's Circuit Stickers are one of the newest ways to combine paper craft and electronics. An engineer and an artist who is finishing her PhD at the MIT Media Lab, Qi was interested in pop-up cards as a form of creative expression, and she wanted to look at ways to incorporate logic and interactivity through electronics. She started Chibitronics with Bunnie Huang to produce "peel and stick" LEDs in the triangular shape of small guitar picks. Circuit stickers are fun and easy to use. The stickers are examples of low-cost flexible circuits, which Bunnie and Qie made in China by developing a new manufacturing process with a factory. They also raised $100,000 through crowd funding to support the development.

The Circuit Sticker Sketchbook kit is a plain brown book that takes a reader through various examples of building circuits and the switches that can be used to activate them. Conductive tape is used instead of wires and is placed along the lines of a diagram. The circuit stickers also work like tape, placed at the edges of the tape, connecting positive and negative. The sketchbook easily suggests how children might build their own book full of interactive pages that illustrate the story and bring it to life. I suspect we may see circuit stickers used in more children's books in the future, upgrading popular sticker books with blinking LEDs. The Circuit Stickers Sketchbook provides an easy and fun entry point for children to gain basic tech literacy in the form of a book, showing creative activities that they can follow on their own.

Ayah Bdeir, who grew up in Beirut, Lebanon, saw herself as a maker when she was growing up, but she found engineering dry and uninteresting, preferring to explore her creative interests. Nonetheless, at her parents' urging, she got a degree in electrical engineering. When she was accepted into the MIT Media Lab, she was able to combine engineering skills and creative work. "I started to create my own artwork using electronics: wearable electronic fashion, interactive installations, lighting art," writes Ayah in the preface to *Getting Started with LittleBits*. "A little while after, I realized I was more interested in the tool than the outcome of what I was creating."[3]

At the Media Lab, Ayah began working on a project that she called LittleBits, a modular electronics system that had the goal of reducing the complexity and the amount of time required to build prototypes. LittleBits is like an updated version of the Braun Lectron kit that inspired Massimo Banzi. Her first prototypes of LittleBits were made out of cardboard and used conductive tape. Eventually, through an iterative design process over three years, she developed the design for LittleBits as modular magnetic connectors called Bits, with different colored Bits for different functions. Blue Bits provide power, such as a battery. Pink Bits represent input, such as an on-off switch. Green represents output, such as a buzzer or an LED. Orange represents wire. A blue Bit and a green Bit are the most basic combination; a battery provides power to an LED, for example. LittleBits is an open-source library of modular connectors that range from simple to sophisticated functions, from lights and sensors to MP3 and Wi-Fi modules. An easy project for a beginner is to create a common household night light by snapping together a power module that connects to a nine-volt battery (blue), a slide switch that can adjust the sensitivity of the light sensor (pink), a sensor that detects light (pink), and an LED output (green).

She brought an early version of LittleBits to Maker Faire Bay Area in 2009 to share with people. Ayah soon noticed the lines of people waiting to check out her electronic building blocks. It helped her realize that there was a market for LittleBits. Ayah found investors, and by 2011, she was able to launch the LittleBits modular library. What started out as a project to build a tool at the MIT Media Lab became a product line and a successful company based in New York and run by Ayah. She has met every challenge that has faced her in growing the company with determination and grace.

Ayah has created an electronics toolkit that can be viewed as a toy by some and as a prototyping system for others. Either way, it succeeds in helping to demystify the technology that surrounds us and demonstrate that it is possible to create new things ourselves using these basic building blocks. "To elevate making into inventing," writes Ayah, "we need to equip ourselves with a new language to understand the world around us, and a platform to reinvent it."[4] She hopes that LittleBits is a platform for more people to learn how to invent the future.

BUILDING ROBOTS

Just as A. S. Gilbert's Erector Set was inspired by the girders hoisted in the air to build skyscrapers in New York City in the early twentieth century, Jasen Wang's Makeblock, which integrates physical and digital modules, is no doubt inspired by makers who are furiously building things in Shenzhen in the twenty-first century. Makeblock is a construction kit that looks like a wholly reimagined Erector Set. It consists of blue aluminum beams, standard mechanical components, and plug-in electronic modules. It features an Arduino-compatible controller and standard sensors. "We combine a lot of new technology, including open-source technology," explained Jasen. It's perfect for building robotic creations that combine electronics and mechanical engineering. Makeblock offers a set of starter kits for building a musical robot or a drawbot, an XY plotter, and even a 3-D printer bot. All of the products are well designed, colorfully packaged, and competitively priced.

Jasen is an engineer, soft-spoken but with a gleam in his eye. Five years ago he came to Shenzhen wanting to start a hardware company after finishing his master's degree. In 2011 he founded Makeblock and was accepted into the first class of companies in the Shenzhen-based incubator HAX. He ran his first Kickstarter campaign in 2013, raising $185,000, becoming one of the first people in China to use Kickstarter to raise funds. More recently, Jasen raised $6 million from Sequoia Capital and expanded his team from ten to ninety people, adding expertise in manufacturing, software development, and design.

At Maker Faire Shenzhen in 2015, Jasen showed me a new product called Mbot, an educational robot for kids. It offers a visual drag-and-drop programming environment modeled after MIT's Scratch to control an Arduino-compatible robot. The Kickstarter campaign for Mbot wrapped up in May, raising $285,000 from 2,500 backers. The kit costs about $75. Jasen believes this robot is "affordable for every kid so they can learn both robotics and programming. Mbot comes in two colors, blue and pink, which will please some and infuriate others. Jasen was able to sell a lot of Mbots during Maker Faire, which he believes has exposed many Chinese parents to the Maker Movement. They realize that making is "important for their kids and themselves."

Makeblock's office is half an hour from central Shenzhen in the Nanshan district, which I was told was a high-tech center. However, what I saw was a set of nine empty high-rise office towers that were developed by the government. Makeblock got a good deal on space in one of the glass towers, a tangible sign that the government is interested in support-ing start-ups and how it expects these companies to grow the economy.

As part of Shenzhen's Maker Week, Makeblock was holding a forty-eight-hour robotics competition at its offices. The theme of the competition was "Lyre, Chess, Calligraphy, and Painting," a prompt to build robots that could play music or chess, paint, or draw. A dozen teams were brought to Shenzhen to compete and provided all the Makeblock components they needed to build a project for the competition. I met teams from the University of Utah and the University of Trento in Italy. Nearing the end of the competition, the projects were nearly finished except for some last-minute testing and battery charging. Several hun-dred people were checking out the projects, and the teams occupied their "pits." Several projects I saw were variations of drawbots—one was a watercolor bot, another splattered paint in creative ways. The winner of the competition was the Light Saber Chessbot from a French team called Dev/null. It was a two-foot-tall chess piece whose movement could be directed by a laser pointer. The teams certainly demonstrated ingenuity and teamwork, and that almost anything can be built with Makeblock.

Like LittleBits, Makeblock can be viewed as a construction toy but also as a cheap and powerful prototyping system, adding mechanical parts along with electronic brains.

BACK TO BASIC COMPUTING

Eben Upton's original idea for Raspberry Pi was to design a really cheap computer that would encourage kids to learn to hack. He didn't want the computer to hide that it was programmable. In fact, he wanted Raspberry Pi to help kids realize that programming computers was fun and powerful. In that sense, Raspberry Pi is intended to be unfinished: just a board that runs Linux. A kid would need to find a monitor and a keyboard and hook it up. A kid would have to make it do something valuable, such as writing code to create his or her own game.

Eben came up with a $25 price for the Raspberry Pi, and then had to work hard to figure out how to build it at that price. "There are 180 components on the Raspberry Pi and only two of them are chips," he said. Those chips were the only components he considered when he first came up with the $25 price. (The basic model sells for $35, but there is now a $5 version.) Eben is also proud that Raspberry Pi is made in Britain, specifically at a Sony plant in South Wales, about ten miles from where he grew up.

In its first year, over a million Raspberry Pis were shipped, which is pretty amazing if not totally unexpected. By 2015 the Raspberry Pi community was celebrating its third birthday at Cambridge University's William Gates Computer Science Laboratory. The two-day event drew about 1,400 people, many of them families with children moving among workshops, talks, demos, and a marketplace of vendors. Of course, being a birthday party, there were balloons, even Mylar ones that spelled Raspberry Pi. Later, there was pizza and raspberry-flavored beer called Irration Ale.

I had a chance to talk with Eben for a few minutes. We were interrupted by a young girl who sweetly asked Eben to autograph her Raspberry Pi board. I asked Eben what he was thinking about Raspberry Pi after three years. "The big thing is how the ten or fifteen people who work for me are focused. I think about how dwarfed we are, how few resources we have. We have to be really good, and my engineers need to stay focused on the core platform, the guts, the stuff we have to do right." Upton believes his handpicked team is exceptional, made up of engineers who are ten times better than most. I asked him what it is in their backgrounds that he looks for that distinguishes them. "It was their hobbyist interest," he replied, immediately and forcefully. "They hacked on computers as kids; they hacked on them at night. This is what they do."

Eben was one of those kids who knew he loved computers at about age ten and started hacking them. He eventually went to Cambridge University in engineering but dropped out in his third year to pursue a start-up. "I knew how to code, but after a while I realized that I wanted to learn the theoretical bits," he said. So he returned to Cambridge to get his degree and eventually his PhD. Now he is back at Cambridge's Computer Laboratory with his Raspberry Pi team and many mostly local contributors to the Raspberry Pi worldwide ecosystem.

To understand the impact of Raspberry Pi, Eben said, "Look at the kids." He had an insight about kids at his first Maker Faire in New York in 2011: "I saw how many kids were lined up to talk to us." He particular remembers a young maker nearby who was exhibiting an Arduino-controlled dollhouse, Andrew Katz.

Eben is very concerned that developed countries like the United Kingdom and the United States are not producing enough of their own engineers. Instead, countries such as China and India produce them for us. "We must do a better job of developing our own talent," he cautioned. This implies also getting more people involved. "Certainly, we want to see more girls get into coding, but look at it this way: we also have only about five percent of boys," he said. "That's a rounding error." He believes that more people should learn to code. "Having ideas is not a contribution. *Implementing* ideas is how you make a real contribution."

If one needed further proof of the impact of Raspberry Pi on kids, a teacher named Sway Humphries offered it. She led a session that featured four of her students, who were around ten years old. Sway said she learned about Raspberry Pi and thought it would be good to use in her school. When she requested that her school buy them for her class, she was turned down "because they thought Raspberry Pi was too new and not yet proven." Undaunted, she brought in her own and began working with the kids. Word began to spread with other children asking her what this "cherry pie" was. Through her participation in an Hour of Code challenge, she won a pack of Raspberry Pis for her class, and that got things started. "I gave them to the kids and said, 'Have a go. See what you can do,'" she explained. They began building their computers, and just getting them up and running was a proud achievement shared by the group. She called her students "digital leaders."

It was interesting to hear the kids talk about how they felt about computers. One said, "The first time I saw a Raspberry Pi, I thought—that's not a computer." In school, they were being taught that using Word and PowerPoint was what it meant to use computers. "Once you learned how to use those programs," said one of the students, "there's not much more to learn." Another said that the Raspberry Pi was "fun to use and hard to use." None of them felt that they understood computers until they began working with a Raspberry Pi. One said that he didn't know how to code, but he was learning, and he could see that it was possible to do many things with it such as math, music, and games. That led to the following prize-winning remark by one of the ten-year-olds: "People who think we are too young to code are wrong."

In another session, Paul Atkin of Cambridge Consultants explained the process of developing remote, real-time photographic stations for use in studying penguins in Antarctica and catching poachers of black rhinos in Africa. He walked through the list of requirements that one of his clients presented him, how he had to come up with a solution that could withstand temperatures to minus forty degrees, be able to take color photos at night and by day, and remain in the field for months running on batteries. Paul discussed how he could use the Iridium satellites in orbit to transmit photos in real time back to researchers. In the case of penguins, researchers would have had to install a camera, hope that it operated successfully for many months, and then return months later to retrieve the images. Now a Raspberry Pi–based solution packaged in a rugged enclosure would allow researchers back in England to see photographs on the same day they were taken in Antarctica. Paul said that it made no sense to develop a custom board when the Raspberry Pi was so capable—and inexpensive.

Raspberry Pi has become not just a cheap computer for kids to learn programming, but also, somewhat unexpectedly, a maker platform for a wide range of projects. Like many of the examples in this chapter, the best toys become tools for learning how to interact with the world. The applications are no longer just toys.

LIFELONG KINDERGARTEN

Mitch Resnick runs the Lifelong Kindergarten program at MIT Media Lab that explores the appropriate role of technology in learning. Mitch developed the Computer Clubhouse in the 1990s, and more recently, the Scratch programming environment for children. Resnick is tall, bearded, and eloquent, speaking just above a whisper. He looks nothing like what I think of as a kindergarten teacher, but that's how he thinks. He is thinking not just about four- or five-year-olds but about everyone.

In a paper, Mitch asked the *Sesame Street* question: "Which of these things is not like the other: computer, television, finger painting?" He believes the nonobvious answer is television. "Until we start to think of computers more like finger paint and less like television," he writes, "computers will not live up to their full potential."[5] Neither will our children.

Mitch was a student of Seymour Papert at MIT in the 1980s. Along with his grad students, Mitch developed the Scratch programming environment for children. Scratch allows children to learn to program using a graphical interface. They can build interactive stories, games, and animations. A very simple Scratch program might move an icon or character around the screen, spinning around in a loop and generating sounds. Scratch can be used as well to interact with physical objects so that it can be a platform for tinkering with hardware and software. Scratch is an online community where children can find code examples created by other kids that they can easily copy and modify. It is a coding playground where kids can play by themselves or in small groups. It's fun, not a chore.

I spoke at Scratch Conference at MIT in 2014. It attracted educators, parents, and museum leaders. Bill and Amanda from Wilkes-Barre, Pennsylvania, came with their two sons, who are homeschooled. They were *Make:* magazine readers as a family, and they were trying to reinvent education in their own way. It's not something everybody can do, but spending several days at MIT Media with hundreds of others and learning what they are doing is about as good an educational experience as one can imagine.

Jay Silver was a student of Mitch Resnick. Jay is the co-inventor of MaKey MaKey, along with Eric Rosenbaum. It is a small red-and-white board that connects through USB to any computer, and then, using alligator clips, you can connect things to control the computer. It is part computer, part toy, and more like finger painting than television.

MaKey MaKey changes the interface for computing to almost anything you want it to be. With MaKey MaKey, you don't have to use a keyboard for your computer, so you interact with it in new, creative ways. Indeed, you can use bananas and apples to play a piano on your computer. Any number of YouTube videos show the wide range of amazing applications that MaKey MaKey enables. The package for MaKey MaKey contains a label:

> WARNING: User may start to believe that they can change the way the world works. Extended use may result in creative confidence.

That you can create a banana piano with MaKey MaKey is a seemingly silly thing, but it is also surprisingly important. Jay calls MaKey MaKey an "invention kit," a new kind of toy or game for "the simultaneous combination of exploration and creative action that leads to a new way of seeing the world." He believes that everyone should learn a new "invention literacy."

I sometimes wonder if young kids today have a "tactile deficit syndrome." It is as though children's sense of touch is underdeveloped. They are growing up as the generation of the touchscreen, accustomed to its glassy look and feel. While a touchscreen responds to a finger, it provides almost no sensation back. The iPad is more like a remote control: the child is controlling a device that engages eyes and ears alone, much like television. Jay wrote to me:

> I really believe in the power of touch. My mom is actually a lactation consultant and has taught me how important skin-to-skin contact is between mother and baby. I have since measured the electrical characteristics of large groups of people holding hands, and of people touching plants and touching the ground.
>
> What I call skin-to-nature interaction, which is most easily implemented with a MaKey MaKey, has special meaning to me. I couldn't believe it when I tried hooking a MaKey MaKey from the leaf of one tree, all the way through the roots of the tree, into the ground, and up through the capillary system of another tree, into the tips of the branches and into the leaves

of the second tree. They are connected through a long circuit: tree, roots, ground, second roots, second tree.

When you attach the alligator clip of the MaKey MaKey to the tree, it closes the circuit. Jay continued, "I once showed this to a kid while experimenting in Boston Commons and his mom e-mailed me the next day thanking me for teaching her son that 'everything is connected.'" Indeed, every living thing that touches the earth is literally electrically connected, except on very dry ground. Jay explained:

> Once I made an interactive jacket called "ok2touch" that I exhibited at the Exploratorium and at the Boston Museum of Science. Skin-to-skin contact with the wearer turned it on. It was a commentary on our modern society's lack of touch.
>
> I suffered from fear of touching nature during my childhood because it was "dirty" or "dangerous." My wife, on the other hand, touches everything she sees in nature, and I've learned the wonder of the feel of moss, cat noses, and caterpillar legs. Touch is so important. And I did design MaKey MaKey quite literally on purpose to "force" people to touch everyday objects, nature, and each other. To give importance to skin-to-skin and skin-to-nature contact as an on switch or a trigger as a part of our fancy interactive technology.

Toys and tools created by makers introduce children to technology, and they also provide an alternative, maybe even an antidote, to screen time. Hands-on play improves sense perception and contributes to an awareness of our own body interacting with nature and the material world. If you can't find the right toy for a child, remember that you have all kinds of materials around the house with which you can make things. Save the cardboard boxes from the things you buy.

GENDER AND TOYS

The significant difference in the number of men versus women who identify as makers, as well as work in technical professions, must be attributable in some measure to the toys that were made for boys, which in turn gave boys the opportunity to develop the skills and self-image of themselves as builders and doers. Perhaps we have Gilbert and his Erector Set to blame in that he and others proved to be successful in the social mission to direct the energy of boys in ways that helped them develop as productive workers in the industrial economy. What worked for boys came at the cost of excluding girls. Today, the social mission of more and more toys is to encourage more girls to develop much the same skill set for a creative economy. New toys aimed at girls or at boys and girls can help introduce them to science and engineering as well as art and design outside school or even before they go to school.

Elizabeth Sweet, a sociologist and lecturer at the University of California, Davis, gave a TEDx talk in 2015 on the gender-segregation of toys in blue and pink aisles in toy stores. The toys themselves embed cultural norms for gender. "Toys for girls center heavily on ideas around beauty, nurturing, and domesticity," she said. "Toys for boys are much more about action, aggression, and excitement." She recalled playing with Tinker Toys, Lego, and Lincoln Logs as a child without thinking they were toys for girls or boys. From her own research, which entailed studying thousands of toy advertisements in Sears Catalogs throughout the twentieth century, she concluded that today, "toys are far more gendered than they ever were at any point during the twentieth century."[6]

It's not that gender was absent in advertising, and she shows an 1912 ad for Gilbert's Erector Set clearly aimed at boys, along for one from the 1960s that featured a girl playing with an iron and ironing board. She found more examples of ads featuring boys and girls playing with the same toy. Yet she believes that the gender stereotyping found in toy stores today is worse, as consumer culture has exacerbated the problem. Its implication that boys can do certain things and girls do different things based on what toys they play with drives gender inequality in our society. We see it in the workplace where women are underrepresented in science and technology professions and men are underrepresented in the caring professions. When there are blue and pink aisles, even toys that are gender-neutral have no place to be. Sweet said that she more finds science kits in the blue aisle. Crossing from one aisle to the other is problematic, as it was for Sweet's daughter, who found a green dinosaur-themed lunch box she liked, but had second thoughts when she read a tag that labeled it a "boy's lunchbox."

Gender stereotyping of toys hurts both boys and girls. Sweet said that these stereotypes tells kids to "abandon or deny parts of themselves that don't fit" the stereotype, and perhaps exaggerate the importance of other parts that they care less about. These stereotypes narrow instead of broaden what is possible for children. Kids should be encouraged to "follow their interests, a diversity of interests" and not be limited to what fits within the bounds of traditional male and female roles. It is the diversity of interests and their intersections that breed creativity.

Sweet believes that the products themselves as well as the marketing can be more gender-balanced. She sees that a toy store without blue and pink aisles would offer children more options, a heterogeneous mix of toys. There could be an alternative world of toys that appeal to boys *and* girls based on their interests.

Seeking gender balance in toys and breaking away from prevalent stereotypes are a big challenge that the new makers of toys are tackling head-on. Anne Mayoral, an industrial designer and codeveloper of the SpinBot kit, wrote in *Make:* "As we design, make and buy STEAM (science, technology, engineering, art, and math) toys, we should think about the message they send to children. Toys and activities that support types of play that defy gender stereotypes will teach the skills, experiences, and intuition that foster an aptitude for STEAM fields."[7] That there are women creating many of these products is reason enough to believe they will drive change in both toys and toy stores.

Ayah Bdeir's LittleBits was intentionally designed to avoid gender-based cues in its product design and marketing, which she admits is very hard to do, and LittleBits has boys and girls as users in the same numbers. Alice Taylor doesn't see the customized dolls that her users create as limited only to girls. The potential for combining paper craft and electronics, as shown by Jie Qi's Circuit Stickers, as well as electronics and fashion in wearables are promising ways to engage girls. New toys such as GoldieBlox and Roominate aim to reach girls and introduce them to science and technology through play. Mayoral writes that the goal for parents should be "to offer a variety of toys and a chance for hands-on discovery and making, either at home or in a local makerspace." Let girls and boys explore all their interests, support their natural curiosity, and encourage them to imagine all that they can make and do.

7

The Maker Mindset

One of the most profound experiences any of us can have is to act on an idea, turn it into something real, and share it. It can be called the act of creation or invention, inspired by our experiences and our imagination and informed by our knowledge and skills. The process of realizing an idea and making it tangible is what defines a maker.

Our own experience as a creator, a maker, a producer can change the world in small but significant ways, and we may not realize it at the time. It can also profoundly change how we think about ourselves, and that kind of change may be the most profound. We develop a sense that our ideas matter, that they can impact us and the world around us. The impact may simply be that we get a person to laugh with us, but that counts for a lot.

Psychologist Mihaly Csikszentmihalyi describes the creative state of mind in which we "want to pursue whatever (we) are doing for its own sake" as "flow."[1] Being in that state is an optimal experience; it makes us happy. That's why the process of making is its own reward, and that's why it is so personal.

As makers, we enjoy repeatedly experiencing this creative process. By taking our own ideas seriously, sharing them with others, and developing them, we give our life meaning and purpose. We have a sense of being in control and having the freedom to choose what to do, and to do things without fear of failure or judgment. We gain confidence.

Engaging in this process develops the maker mindset. When I talk about the maker mindset, I mean all the intangibles that come as a result of the very tangible experiences of making. Makers acquire this mindset through the practice of making. It's not necessarily intentionally sought out on its own—it develops with experience.

What are the qualities of the maker mindset? Makers are active, engaged, playful, and resourceful. They have a well-developed sense of curiosity and wonder. Makers are self-directed learners, able to figure out one way or another how to learn what they need to know. They learn to use tools and technology to create new things. They are willing to take risks, trying to do something that others have not done or creating something that they have not seen before. They are persistent, overcoming frustration, and resilient, trying again when they experience failure. Makers are resourceful, developing the ability to make do with what is available or exploring alternatives that might be cheaper or better for the environment. Makers are good at improvising: they are able to do things that have no instructions. Makers are generally open and generous, willing to share their work and their expertise, often helping others in the recognition that they have benefited themselves from such help. Makers believe in their own individual agency to act and create change in their own lives and their community.

I meet makers who have widely different interests or live in very different places and work under different conditions, yet they share the same mindset. Having this mindset in common allows makers to connect easily with each other as though they had known each other for a long time. This mindset opens doors to new opportunities for personal and social development.

Carol Dweck, a Stanford psychology professor, wrote a book called *Mindset: The New Psychology of Success.* Dweck's twenty years of research show that "the view that you adopt for yourself affects the way you lead your life."[2] It seems obvious, but how aware is each of us of the mindset that we have and whether, in any meaningful sense, we adopted it? Can you remember the moment when you adopted your mindset? I suspect many of us think that our mindset was something we had from birth, just like our appearance and personality. Dweck implies that a mindset is trained: while genes undoubtedly play a role, they are only a starting point. We can think of a mindset as something that "you can cultivate through your efforts."[3] We aren't stamped at birth with a mindset; yet what we believe about ourselves changes how we live our life.

Dweck distinguishes between fixed and growth mindsets. A fixed mindset reflects the belief that one's capabilities have already been determined, and developing new abilities is not possible. A growth mindset reflects the belief that one's capabilities can be developed, improved, and expanded. When asked to try something new outside his or her comfort zone, a person with a fixed mindset is more likely to decline, thinking that there's only downside and nothing to be gained. People with a growth mindset are more likely to embrace the opportunity readily, without thinking about whether or not they will be successful. People with these different mindsets exhibit different attitudes toward risk and potential failure.

Dweck points out that many people who excel academically in school have a fixed mindset, which limits them to exploring only the areas for which they believe they have an aptitude. Said another way, they stick to doing what they've been told they are good at. Particularly in the past, a fixed , through its very limitations and predictability, was often a path to success. However, it's a not a path that leads to creativity or innovation. A growth supports the belief that we can develop and change, especially by learning new things.

In a world that grows more interconnected and interdisciplinary every day, a growth is a fundamental advantage for us to adapt to change, if not become an agent of change. Moreover, a growth predisposes us to believing that our own actions matter and that we can change the world instead of accepting the status quo.

The maker is an expression of the growth that is evident in a maker's willingness to learn new tools and methods as well as experiment without certainty of success. Because of this , makers are optimistic about what they can do. I see it and feel it present in makers at a Maker Faire—and it comes from people who are fully engaged, doing something they love to do, and believing that what they do is worth sharing.

THE EFFORT-DRIVEN REWARDS CIRCUIT

Is there something special about making and using our hands that might support the development of this kind of? Kelly Lambert, a professor of psychology at Randolph-Macon College who runs a behavioral neuroscience lab, began exploring hands-on activities as way of relieving the symptoms of depression. Lambert wrote about her research in the book, *Lifting Depression*. She sent me a copy after seeing me on CNN.

Her interest in depression began following her mother's death with her own sadness and depression. After weeks of feeling that none of her "efforts made any difference in the world," she found relief in vacuuming her house, something she normally did not like to do. The physical work made her feel better. "Each time I saw tangible evidence of the dirt and grime I'd physically removed from my house, I felt my efforts were valuable." It gave her a sense of control over her environment. That experience led Lambert to explore the neuroscience behind depression in her lab, as well as the correlation between hands-on work and how we feel about ourselves.

> What I've discovered is that there's a critical link between the symptoms of depression and key areas of the brain involved with motivation, pleasure, movement, and thought. Because these brain areas communicate back and forth, they are considered a circuit, one of many in our brains. In fact, the rich interactions along this particular brain circuit, which I called the effort-driven rewards circuit, provide us with surprising insights into how depression is both activated and alleviated.[4]

I love that she calls it a circuit, something makers can understand. When all the parts are linked together properly, there's a flow of energy through this circuit. We feel engaged by our actions, alive in our minds, and interact easily with others. Our brain is giving us positive feedback that is a result of our effort. When the circuit is disengaged, we feel blue, as though it wouldn't matter what we do. Lambert writes:

What revs up the crucial effort-driven rewards circuit, the fuel, is generated by doing certain types of physical activities, especially ones that involve your hands. It's important that these actions produce a result you can see, feel, and touch, such as knitting a sweater or tending a garden. Such actions and their associated thoughts, plans, and ultimate results change the physiology and chemical makeup of the effort-driven rewards circuit in an energized way. I call the emotional sense of well-being that results effort-driven rewards.[5]

Lambert can't emphasize enough how central the hands are to this circuit. "Our hands are so important that moving them activates larger areas of the brain's central cortex than moving much larger parts of our bodies, such as our back or even our legs." Our hands are uniquely connected to our brain, and hand movements are "the most effective way to kick-start the circuit in to gear." This runs counter to our usual separation of manual and mental labor, of physical and mental, of hand and mind. What if the phrase "hands-on" were to be associated in our minds with a heightened mental state? We know of people who talk with their hands, but makers are people who think and communicate with their hands.

The harder the work, the more rewarding it is. It is a fine line: if it too easy, there is little reward. Yet if it is too difficult, we will just give up. Experiments that Lambert did in her lab with rats led her to the conclusion that persistence can be learned. It may be that the sustained effort is what matters, not simply exertion. Prolonged efforts to make things are deep experiences that not only activate our brain, but change it, initiating growth, creating new connections. Actions that we see as meaningful "likely stimulates neurogenesis—the production of new brain cells,"[6] writes Lambert. We arc changing our minds as we change the physical world around us.

The symptoms that Lambert associates with depression—loss of meaning, loss of pleasure, sluggishness, poor concentration, slow motor responses—might be considered the opposite of the maker : purpose, joy, engagement, focus and flow, and resilience. The maker could be the product of repeatedly engaging the effort-driven rewards circuit with activities that use our hands as well as our brain. If you enjoy making, you'll do more of it—your brain tells you it wants more. Our bodies and our minds like to play well together.

Lambert thinks that the rise of depression in our culture could be tied to "effort*less*-driven rewards," a consumer culture that provides rewards with greater and greater convenience so that there's little physical or mental work associated with getting them. It's what fast food is: the "reward" is a bunch of calories that cost us little money and little time or effort. We take less pleasure in the food than we might have had we spent the time preparing it. Lambert cites additional research that shows that if the effort-driven rewards circuit is disrupted, instead of giving full effort for maximum reward, we can learn to settle for a smaller reward that requires less effort. We become complacent.

In my view, the effort-driven rewards circuit gives us a model for understanding how making in itself produces the kind of physical and mental well-being that we find in the maker —why making makes us feel good.

PLAYFULNESS

If I had to give a prescription for the maker , I might say: "Be more playful." It really doesn't matter what you choose to make or how good you are at doing it. What matters most is jumping in and enjoying the experience. This is the practice. The more it feels like play, the more you'll enjoy it.

With *Make:* magazine, I had an insight that adults needed to play and rediscover hobbies and passion projects. I saw that makers liked to play, whether they used that word to describe it or not. Perhaps it was enough to realize that the problem at hand had their full attention and everything else fell away. Makers didn't consider what they were doing to be work, and they didn't necessarily know where it might lead them.

Perhaps the most important thing for adults is that play can be entirely under your own control. You do what you want to do. There are no committees that have to decide, no hierarchy to navigate for approval, no external conditions placed on your own interests. Control is in your own hands.

I like to call what makers do "experimental play," as John Dewey, the educational philosopher, uses the term, meaning that we are testing what we understand and what we can do. Experimental play creates a context for us where it is safe to try things: the stakes are low, judgments are withheld, and there's no prescribed goal or outcome.

Yet my sense from seeing so many tech enthusiasts at play is that it has an additional benefit. Experimental play created the conditions for innovation. That is, the immersion in a set of problems or a set of tools gives rise to new insights that can lead to unexpected solutions and unplanned products. Innovation can emerge from our own set of experiences. If makers did not play with drawbots and broken 3-D printers, they would not have immersed themselves sufficiently to have new ideas about how they could be improved. Through play, makers can see what's missing, what doesn't work as it should, what was poorly designed and needs to be completely rethought. A maker might make the assertion: "I can do better."

In his book *Play: How It Shapes the Brain, Opens the Imagination, and Invigorates the Soul,* psychiatrist Stuart Brown declares "Play lies at the core of creativity and innovation."[7] Brown tells the story of how Caltech's Jet Propulsion Laboratory (JPL) realized that, although they were hiring the best and brightest college graduates, they were the wrong kind of people to create the solutions they needed. Something had changed in the people who came to work there:

> The JPL managers went back to look at their own retiring engineers and … found that in their youth, their older, problem-solving employees had taken apart clocks to see how they worked, or made soapbox derby racers, or built hi-fi stereos, or fixed appliances. The young engineering school graduates who had also done these things, who had played with their hands, were adept at the kinds of problem-solving that management sought. Those who hadn't, generally were not. From that point on, JPL

made questions about applicants "youthful projects and play" a standard part of job interviews.[8]

We might say that JPL realized it was looking for a , and this seemed to develop not in school but through "youthful projects and play." Just having a degree doesn't guarantee that you have the right .

Brown, like Lambert, makes the case that play reshapes our brain. "Play is nature's greatest tool for creating new neural networks and for reconciling cognitive difficulties. The abilities to make new patterns, find the unusual among the common, and spark curiosity and alert observation are all fostered."[9] Both Lambert and Brown, as psychologists, might look at the therapeutic value of play for adults, Brown emphasizes that play is essential for our well-being.

Some people will argue that they don't have the time to play or make. It's an argument Brown has heard as well. We're too busy working to play. Yet, creating time for play is also essential to balance our work lives with our own interests. Brown writes that "the opposite of play is not work—the opposite of play is depression." We need both work and play. Brown notes that they are "mutually supportive," with play providing "a sense of discovery and liveliness,"[10] and work providing us with a sense of purpose and that we are needed by others. In fact, Brown also admits that play and work can merge for us. Often when we are doing our very best work, we feel like we are playing.

In dealing with adults who struggle to have a sense of play, Brown asks them to develop a play history. He asks them to think back to their childhood and recall periods of play, what they were doing, where it took place, and how it made them feel. To adapt the idea from Brown, I think we might also consider composing our own "make history": recalling experiences of building, creating, designing something from scratch, reflecting on what we were doing and how it made us feel. This can help us recall the very personal connection between making and play, and the satisfaction of creating something new. Rediscover something you enjoyed as a child, a hobby or passion that can be rekindled.

THE HANDS-ON IMPERATIVE

"Now you do it," said Mister Jalopy to me. It was my turn, and I was a bit hesitant to take it.

Mister Jalopy, a former music industry executive in Los Angeles who now owns a variety store and various laundromats, was giving several of us a welding demonstration. He considers himself a "mediocre welder," yet he is a "jack-of-all-trades" and seems to able to do anything, just nothing particularly well, he would say. Mister Jalopy explained the welding equipment, a MIG welder, and demonstrated how to put on the helmet, with its auto-darkening mask, along with leathers and thick gloves. He looked really cool. He explained the difference between a good weld and a bad one, and said he had his own record of doing both. Then he said we should each try it ourselves. He warned us: the first time you do it, you might find it hard to lay down a bead. First, he showed us himself. Then he turned to me and said: "Now you do it."

Something in my head said I wasn't ready. I wanted to see others doing it first. I was reluctant to just try it. Mister Jalopy insisted that it was my turn. He handed me the helmet and then I put on the gloves. My heart was racing, and I knew I had to hold the tip of the welder steady. I focused on the tip of the welder, where it was melting the metal, trying to create a continuous line.

"Good," said Mister Jalopy when I was done. I didn't necessarily do a good job, but I had done it. It didn't make me a good welder, but it did demystify welding for me. I had learned to do something that I had never done before, and I was secretly proud of myself, just for trying.

We've all experienced it, this moment of truth, when it is our turn to try to do something we haven't done before. Any number of people could tell us how to do it, but it isn't until we actually do it that we feel we've learned how to do it. It is like first learning to drive a car. We watched other people do it, and then were shown how to do it by a parent or friend. We probably studied a driver's handbook. But when it was our turn to sit behind the wheel, we actually had to do it. We were responsible for the outcome—we were in control. No matter how much we had rehearsed the procedure in our mind, there was a question about what would happen when we actually did it. This moment can be filled with apprehension and self-examination. *What if I don't do it right?* We know what to do, but doing it is another matter. It gives us pause, like standing on a high diving board and looking down at the water. However, this moment of truth demands that we take the leap.

It's the same experience in learning to use a drill press, a sewing machine, or a band saw. Someone who knows the machine shows us how to use it: how to turn it on, how to prepare the job, and how to follow a process to get something done. Then this person looks at us and says: "Now it's your turn." No amount of observing, reading, or thinking quite prepares us for this moment. We don't really believe what we know until we do it ourselves the first time, and then again. No one is very good at the beginning, but through repeated practice, we get better. This is how we develop agency, the sense of having control.

THE PRACTICE OF MAKING

Makers may not necessarily have formal training or follow the conventions of those who do it professionally. Makers are interdisciplinary, moving across the boundaries of fields, even as amateurs. To paraphrase what Shunryu Suzuki wrote about the beginner's mind in Zen: In the maker's mind, there are many possibilities, but in the expert's there are few.[11] Makers are comfortable participating as nonexperts.

Designer Charles Eames said that anyone could become a designer with persistence and time: "Start with a single idea, do 100 iterations of that idea, choose one of those instances, and then iterate 100 times again."[12] His point was that anyone willing to do the work of iterating over an idea would learn everything they need to know about what they were doing and become good at doing it. Iteration is the cycle of repeatedly changing a process or experiment, until better results are yielded or until mastery has been achieved.

Students often learn a formalized design or engineering process, yet I believe it can interfere with them learning from their own experience. Initially, experimenting and improvising—finding one's own way into and through process—will seem messy, especially if it involves learning to master new techniques or tools. However, by gaining expertise through practice and repetition, we begin to gain a sense of our own personal process, which is valuable when new questions and challenges come along.

We don't have to get things right from the start. Rapid prototyping means that we can iterate more often, taking advantage of technology that can make it easier and cheaper to build prototypes. Technology can decrease the time between iterations while also increasing the number of iterations overall. We can take a rough idea and build a rough prototype, then iterate—the more the better.

"It's all about iterations," Massimo Banzi told me as we talked about how he helped students learn interactive design:

> You start with sketches and prototypes. Then you have different fidelities. You might start with cardboard, as an example of a low-fidelity prototype. When you stuff some electronics in there, the prototype starts to behave like the thing you want to design and so this becomes a high-fidelity prototype.
>
> These iterations allow you to try the prototype with people. Since I believe that interaction design is about trying things with people, the more you want to make your product perfect, the more you need to be able to play with the product. This means that the shorter each iteration, the more experiments you can do. The tools that we use are chosen because of their ability to shorten the loop.

Massimo and his team developed Arduino to enable rapid prototyping of projects. This describes an experientially driven design process that I think many makers follow to develop a project. It's very different than what is taught in many design and engineering schools. The formal process is often taught in a way that a large amount of time is spent planning, developing specifications, and doing research before any attempt is made to build. In this model, ideation is cheaper, and iteration happens by talking over the plan, instead of actually incurring the costs of building the thing. The model that Massimo talks about for interaction design requires getting to a physical representation early, and iterating over it often. That's a very basic, organic view of design from a maker's point of view.

Formal models of this process, such as design thinking, which comes from David Kelley and the work of the design firm IDEO, describe a human-centered design process for understanding a problem, focusing on it deeply, then developing new ideas for solutions, building out prototypes and then testing them. The stages can be presented as:

- Empathize
- Define
- Ideate
- Prototype
- Test

Design thinking is a representation of the design process that allows us to talk about it. It's not the process itself. Indeed, the process we experience is not linear and less logical than formal models make it appear.

Gary Donahue, a teacher at an international Pre-K–5 school in South Korea, shared a video with me of his fourth-grade class giving a tour of their makerspace. It begins with a young boy pointing out the elements of the design cycle, consisting of Investigate, Design, Plan, Create, Evaluate. What shines in the video is the children giving a tour of the makerspace and describing the materials available to them. What speaks to me is seeing the students with their creative projects, from robots to windmills to wearables, and their sense of joy and excitement.

Formal teaching can sometimes leave out the joy and excitement, substituting an emphasis on rigor, formulas, and systems that seem like ends in themselves. Typically, engineering students have a lot less choice in what subjects they take and how much time they get to make things. In short, they aren't encouraged to tinker. Some think it doesn't reflect well on engineering as a profession to talk about tinkering. In fact, on a visit to the National Academy of Engineering in Washington, D.C., a leader of a group responsible for engineering education in the elementary schools told me that "engineers aren't makers." I shook my head, waved a copy of *Make:* at him, and said that the next generation of engineers are reading this magazine. The guardians of the engineering profession would like us to think of engineers as doctors in lab coats, not the Maytag repairman. They would like kids to aspire to join the engineering profession from an early age. They don't like people informally calling themselves engineers, without the proper credentials; in fact, there has been an ongoing debate about programmers who call themselves software engineers.

The process of makers might be informal, messy, and organic. It's a process that includes but overcomes repeated failures, misunderstandings, and kludges. It's their own process that reflects real life rather than an ideal or model. A maker thinks: if it works for me, it works. If it doesn't, keep changing it. Also, makers learn from others; they can reflect on their own process, compare it to others, and borrow what they need.

In an interview by William Lidwell in *Make:* volume 4, the inventor Dean Kamen was asked if he saw himself as more similar to Thomas Edison, whose method was trial-and-error on a large scale, or Nikolai Tesla, who worked out his models in his head before ever building them. Kamen replies:

> Unfortunately, I would put myself closer to the Edison end of the continuum: the tinkerer, the get-your-hands-dirty and keep-screwing-with-it-until-you-make-it-work side. I am much more in awe of people like Galileo, Newton, and Einstein than I am of the tinkerers who just kept working with the tools and technology of their day until they got something to work. I am just in awe of those people. I wish I was one of them, but it's not in the cards. So I work hard to succeed at the other end of the scale.[13]

Kamen, a successful self-taught inventor and engineer best known for the Segway, considers himself more a tinkerer than in the ranks of scientists who discovered fundamental truths about the world. Yet Kamen is a genius in his own way, as are others who may call themselves tinkerers or engineers, with or without formal training.

Engineering is undervalued in Western culture, especially compared to science, writes Steven L. Goldman, a distinguished professor in the humanities at Lehigh University, in a paper "Why We Need a Philosophy of Engineering." He writes that engineering is associated "with the probable, the particular, the contextual, and the temporal," while science is associated with "the necessary, the certain, the universal, the constant, and the timeless." Placing science on a level above engineering "subordinates practice, values, emotion, and will to theory, value-neutral principles, and deductive logic in ways that leave us ill-equipped to deal rationally with life."[14] I believe the maker reflects a shift in valuing practice over theory and uncertainty over certainty.

Craig Forest and his colleagues at Georgia Tech, who have supported the Invention Studio for students described in chapter 4, writes: "University environments that foster open-ended design-build projects are uncommon.... In standard engineering curricula, students do not generally create or invent anything tangible until the culminating Capstone Design experience."[15] There was a major shift in engineering departments between 1935 and 1965 that moved away from "hands-on, practice-based curricula." A new curricula was developed that fostered "theory-based approaches with a heavier emphasis on mathematical modeling."

So universities load up students with textbook knowledge, expecting that they will find some of it useful when they find a job after school, and it is on the job that they will get more specific, more practical training. Yet the National Research Council in 2004 said that the engineer of 2020 must be able to create, invent, and innovate, and schools should do more to help students cultivate these skills. That's what the Invention Studio does. When companies come looking for summer interns, the students in the Invention Studio are their first choice because they are "doers."

Perhaps, as Goldman points out, we should understand the ways in which science and mathematics are different from engineering and technology, despite the convenience of an acronym like STEM that packs them all together. "Engineers use mathematical and scientific knowledge to solve their problems," writes Goldman, "but they do so in ways utterly different from the ways that mathematicians and scientists solve their problems." An engineer like Kamen is looking at solving different problems than a scientist like Einstein, so their methods are different. "Where scientists aim at the truth about nature," writes Goldman, "engineering design reflects ... consciously operating under conditions of partial information and acting on solutions judged good enough to do the job that needs to be done, even though they are not optimal."[16]

Engineering is pragmatic, especially as practiced by makers. Makers also understand that they aren't perfect and they have gaps in their knowledge or abilities. Yet that doesn't keep them from making.

Ugo Conti is the seventy-year-old designer and builder of a prototype of a whole new kind of boat called Proteus. It suspends the body of the boat above the water on spindly legs that make it look like a water spider. Ugo is Italian but moved to the United States to get a PhD from the University of California, Berkeley, and never left the Bay Area. He works in a garage shop, building much of his boat there. If he wasn't working on his boat, he was working on his house, crafting a beautiful wooden spiral staircase for his home. He is wonderfully Italian, combining artistry with "a feel for engineering," the title I gave to the article about him in *Make:*. What fascinated me about Ugo was his emphasis on the gift of intuition:

> I was born an engineer. That's what I am. The instinct of wanting to understand how things work and using that understanding to do something, to make something—in my case that started at a very early age. I have intuition about how things work. I understand certain simple things, like the laws of physics, for instance, without mathematics. I'm not a mathematician. I don't do mathematics.
>
> I'm a very normal person, but I have big peaks. It impresses people, because I can approach a problem without knowing anything about it and come up with a solution. It may be a problem people have been working on for months, but I solve it quickly, just out of intuition. It's a gift. I was born with it. People look at the peaks and think I'm a genius. Well, yes, in the peaks I am, but most of the time I'm just normal. In fact, I make a lot of mistakes. I mean, one mistake after the other.[17]

The philosopher of science Karl Popper wrote, "All life is problem solving." He added that "all organisms are inventors and technicians, good or not so good, successful or not so successful, in solving technical problems."[18] What to eat? What to build? Where is it safe to live? All organisms, including people, have to figure out what to do each day to get food and find a place to sleep, and even more, what makes us happy. If we accept what Popper says, then a person who is not solving problems is not really living, and an education that does not solve real problems is not really about learning, and a job that does not have us solve problems is not really working.

Emily Pilloton liked problem-solving as kid. She is the founder of Project H, which was the subject of a documentary titled *If You Build It*. In an interview by Stett Holbrook in *Make:* volume 40, she said that what excited her as a kid was "the kind of MacGyver-style eagerness of solving a problem under tight constraints." She adds: "I love being constrained—having $10, one hand tied behind my back, and being blindfolded, having nothing and making something beautiful out of that." That attitude is reflected in what she has done working with students, first a group of high school students in Greenville, North Carolina, and now an all-girls class in Berkeley, California. Emily's goal is to "excite young people in a way—inside or outside of school—that is meaningful to them, that is meaningful to their communities, and that helps them bring their ideas to life to ways that maybe they didn't think possible."[19] As they develop as makers, they will develop a maker .

Makers enjoy solving problems, so they are willing to spend lots of time trying to solve them. Some people would ignore the problem or find someone else to solve it. However, a maker not only wants to solve the problem but believes that there's some intangible benefit gained from the experience alone.

Initially, the problems may be small and personal, but makers are also tackling bigger, more ambitious problems as they get good at prob- lem-solving. And they are doing it by collaborating openly on open, distributed projects such as e-NABLE, a global network of volunteers who are using 3-D printers to create prosthetic hands for children. They are sharing and modifying open-source 3-D designs but also educating people on 3-D printing. At Maker Faire Bay Area in 2015, I met a father from Cincinnati whose son who was showing off his red 3-D-printed hand. The father told me that he knew nothing about 3-D printing just two years ago, but then he learned about this community and realized that he could get involved and help his son.

A riff on the Chinese proverb "May you live in interesting times" might be "May you find interesting problems." Makers are defined by the problems that interest them the most.

OPENNESS

"As we enjoy great advantages from the inventions of others, we should be glad of an opportunity to serve others by any invention of ours," wrote Benjamin Franklin in his *Autobiography*. "This we should do freely and generously." In 1742, Franklin had invented a new stove, first called the Pennsylvania fireplace but later more commonly known as the Franklin stove. It was more efficient at generating heat from less wood and did a better job of circulating heated air in a room. The governor was so pleased with the construction of the stove that he offered him a ten-year patent. Franklin declined it "from a principle which has ever weighed with me on such occasions." Franklin published a pamphlet about his stove, and others began making it, at a profit. Yet Franklin preferred sharing his work openly, as he said, "freely and generously." What troubled him, and it remains a problem with the patent system today, was that by choosing not to obtain a patent, he left the door open for others to obtain a patent that could potentially restrict the free use of his own inventions.

Franklin's thoughts are reflected today in the maker community and its development of open-source hardware. Makers are influenced by the notion that "information should be free," which was embedded in the culture of the Internet as well as Richard Stallman's Free Software movement. As a result, the maker community started out with sharing as a default setting. It became an accepted way of doing things because of its very personal benefits. However, it would be wrong to think of openness as a requirement. It's a choice made by the maker balancing his or her own interests with the best interests of the community. Makers usually realize that choosing to share is a kind of payback in return for the benefit they received from others who shared their work.

Open-source hardware is a formal expression of the responsibilities of those who share their work and those who make use of that work. Openness is more than open source, more than a license. It is part of an open, collaborative culture that supports art and innovation, not just in individual or business terms. It encourages participation without asking for permission. It allows for work to be passed around and passed on without limits.

TRUE BELIEVERS

In my office I have a picture of a lettered roadside sign, supposedly created by a student at the Rhode Island School of Design, that reads: "All I want to be is someone who makes new things and thinks about them." That "all I want to be" reflects the zeal of an inspired maker. It is more than a mental state. It integrates our whole physical, mental, and emotional being.

The origin of the word *enthusiasm* is Greek, meaning "the god within." Perhaps in our context, it could mean "the maker within." To be enthusiastic means to be inspired, eager, rapt, or in ecstasy. It's associated with intense emotion. Enthusiasm describes an energy or power within us that we discover in ourselves, and it fills us up. The fascinating thing about enthusiasm is that it seems to be something we use up, but doing so seems to generate more of it. We don't lose it by using it. It's a kind of eternal flame. When I see enthusiasm in a person, I trust that what they are doing is done with conviction. I know they are true and not false. I don't look for other motives in why they do what they do. I trust that they are both passionate and dedicated. That enthusiasm was what made me first notice makers, partly because I had learned to recognize it in myself.

"We are agents of creative change, but only if we see ourselves that way," said Jay Silver at Maker Faire Bay Area in 2013. The coinventor of MaKey MaKey, described in chapter 6, retains the sense of joy and wonder that we see in children and that we ourselves experience when we look at the world with fresh eyes. At a talk at Maker Faire Orlando, Jay told a story about how his young son discovered that flying gnats responded to the sound of his voice, and together they hooked up a speaker to see the gnats flying in a pattern set by the sound waves.

Jay says, "We don't want everybody the same. We want a diverse ecosystem of unique thinkers. The Maker Movement isn't about robots or 3-D printers almost at all, it's really about freedom, the freedom for us with our hands to make the world we live in."[20] Jay now has a PhD from MIT, having completed his thesis, *The World as a Construction Kit*. The world is reconfigurable: Everything in the world can be used to remake the world. It is right in front of us if we choose to see it that way. That kind of worldview is fundamental to the maker . The things around us are not ends in themselves, but components or parts that can be remixed in projects of our own desire.

At Maker Faire Oslo in 2014, I met Erik Thorstensson of Göteborg, Sweden. Erik developed a simple, modular system for building with plastic straws called Strawbees. An open-source design, Strawbees uses simple connectors that can be die-cut on demand from scrap plastic or bought as a kit. With the simplest of materials, it teaches how to build 3-D objects, even complex ones. I watched young kids and adults walk up and build structures from straws: a pointy hat, a wand, a diamond that folds in on itself, a pyramid. All of them were having fun, and Erik was the ringleader, usually accompanied by a team of like-minded instigators. He wears a bow tie, adding a touch of formality to his flair for showmanship. He has boundless energy, talking to everyone and encouraging them to get started creating something with Strawbees. He is always adding new things and expanding his system. In Singapore, he used Strawbees to build a winged aircraft. Adding a few motors and a brain, he turned it into a drone and flew it out over the audience. Erik puts heart and soul into what he is doing.

Erik is a true believer, a phrase Mister Jalopy once used in a talk. Mister Jalopy meant that makers are true believers who act out of the belief that what they are doing is worthwhile. They are committed, not sitting on the fence. Makers believe in their work long before they know it will succeed. Makers believe in engaging others in their work, spreading it as far as they can.

Making fulfills a deep human need to create. The poet Frank Bidart writes:

> *There is something missing in our definition, vision, of a human being: the need to make.*
> *We are creatures who need to make....*
> *Making is the mirror in which we see ourselves.*[21]

That human need becomes a belief system that supports our understanding of the value of what we create and why. We create meaning in our lives by discovering the maker within us.

THE AMERICAN MAKER

The maker can be found as an element of any nation or culture, although it may be expressed differently in each one. It is not a uniquely American trait, although our national pride makes us think so. Because we can find it in the past, we think of making and its as part of the American character.

Walter Isaacson, the biographer of Benjamin Franklin and Steve Jobs, wrote a piece in the *Washington Post* titled "The America Ben Franklin Saw," in which he commented that the American character has two separate strands: one a "liberty-loving individualist" and the other a "civic-minded citizen who sees the nation's progress as a common endeavor." He sees Franklin as "the first embodiment of that American archetype." Franklin "believed that the business of America was not merely to celebrate success but also to ensure that each new generation had the opportunity to achieve it,"[22] wrote Isaacson.

Franklin might be the first famous maker in America. He is practical but endlessly curious. He is the self-made man, starting out as an apprentice, lacking much formal schooling and learning from real-world experiences. He ventured into any area that interested him and made significant contributions to different fields, while also doing well in business and politics. His many practical inventions in addition to the Franklin stove are bifocals, lightning rods, and swim fins. His kite experiment famously allowed him to discover and describe important properties of electricity, coining terms such as *battery* and *conductor* that we use today. The lightning rod, which saved wooden buildings from burning to the ground, was a practical application of what he had learned about harnessing electricity.

Thomas Jefferson's home at Monticello, Virginia, is a must-see museum of an American maker who also pursued many interests, influenced by the Enlightenment. On a tour of Monticello, I saw the writing instrument, an invention called the polygraph, that Jefferson used to create copies of his correspondence. The polygraph has a second writing pad whose pen is mechanically linked to the pen used by the writer, so it composes a duplicate copy. Jefferson didn't invent the polygraph; he was an enthusiast and early adopter, trying it out and communicating with the inventor and builder about how it might be improved. Given that he wrote twenty thousand letters in his lifetime, one wonders what Jefferson would have done with e-mail.

Yet there were many more American makers we don't know by name but who shared this : men and women alike who, out of the necessity of having to provide for themselves, acquired practical skills. The Foxfire Book series, first published in 1970s, developed as a project led by rural high school students to document the how-to knowledge of previous generations living in the Appalachian Mountains when they realized that very little of it had been written down. The first book in the series covers basket making, soap making, building a log cabin, slaughtering a hog, and "moonshining as a fine art." It is a handbook on the ways in which pioneers struggled to be self-sufficient, living off the land.

What makes Foxfire special is that the students rediscovered this knowledge for themselves and learned to appreciate their past. "It wasn't until I had worked on Foxfire for five months," writes a student, Paul Gillespie, "that an inexplicable void between myself and the old people of our region disappeared."[23] He went to visit Aunt Arie, who lived in a remote area by herself. "I was apprehensive because I didn't know what to expect," writes Gillespie. He thought of her log cabin as a time machine that took him back to the 1880s. "Everything she had—from the stern-looking pictures of her grandparents to the fireplace that was her only source of heat—made me stop and look deeply for the first time." Aunt Arie wasn't just sitting in her rocker when the students arrived to interview her. She was, with some difficulty, removing the eyes from a freshly killed hog's head. She asked them to help, and told them that she was making souse meat, "the best stuff I ever eat."

The Foxfire books can still shock us today into seeing what life was once like without all the conveniences. "We made a good life here," said Aunt Arie, "but we put in lots'a time. Many an' many a night I've been workin' when two o'clock come in th' mornin'—cardin' 'n' spinnin' 'n' sewin'."[24] It was constant hard work, but one can appreciate how much they knew how to do.

"Our ancestors brought little more than a few tools and a great deal of ingenuity," the editors of Foxfire write. "They had to find ways to convert wood into most of what they needed for survival, and the resulting reverence for and skill with wood was boundless and profound."[25] Sounding like a maker from a fab lab, they say that so many tools used for woodworking were themselves made of wood. They had a very specific vocabulary for the skills, tools, materials, and techniques used to transform wood in useful ways. Almost all of that knowledge was transmitted orally from generation to generation as lore. Some of this knowledge was written down and published, which allowed it to be shared more broadly.

In Napa Valley, amidst the vineyards, there's a historic site called the Bale Grist Mill that stands as a reminder that the valley was first known for growing wheat, not grapes. The mill was built in 1846, just before California was to become its own republic, independent of Mexico, and before the gold rush. An English doctor who had gained a bad reputation in the Mexican Army, Edward Turner Bale got a grant of land in Napa and built a sawmill and a gristmill.

I visited the restored gristmill on a summer day when it was in operation. The most prominent feature of the gristmill is a thirty-six-foot overshot waterwheel. Water comes from a nearby pond and is directed by a wooden chute down a flume on top of the wheel to make it turn. Here is an elemental machine, powered by water and made of iron, wood, and stone. When it starts up, you hear the water begin to fall, and the waterwheel creaks a bit and begins to move. As the large wheel spins and the wooden gears kick in, the pair of grindstones begin to spin against each other. The whole building rumbles as kernels of wheat pass between the stones, and out of a chute comes a soft white powder.

The grindstones are large wheels cut from granite. The idiom "nose to the grindstone" comes from the constant attention required of the miller to check for ozone, which is caused when the two large pieces of stone are rubbing against each other. It has come to mean paying close attention to one's work, but originally involved using one's nose to smell the functioning of a machine.

I wondered how such a mill would be built in the 1840s. The construction required the work of different trades that weren't likely to be found in the area. Perhaps one might hire an itinerant team experienced at designing and building a mill. The point is that you couldn't buy a mill and have it delivered to you. If you wanted to own and operate a mill, you'd have to build it. How big a waterwheel do you need? What size gears? What is the best proportion for all the various gears?

The answers weren't to be found in the local community but rather in a book, first published in 1795, called *The Young Mill-Wright and Miller's Guide* by Oliver Evans. It was an early American user manual for the design, construction, and operation of a mill. Evans was an American inventor, and a biography of him says that he was known for two things: having a "compulsive urge to invent" and a zeal "to publish technical information for the guidance of young men."[26] His guide for millwrights, which was both theoretical and practical, incorporated his own inventions and ways that his American-made mill was an improvement on the English mill. He also pointed out areas where new inventions were needed.

A fifteenth edition of the book, published in 1860, can be found on Google Books.[27] It's a rather strange how-to book, but it begins with chapters on the theory of mechanics. He explained how and why mills work in a very detailed fashion for a person without any previous knowledge. It was the one book you'd want if you were building a mill in a community that didn't have millwrights. It allowed ordinary people with perhaps some skill as in carpentry to build a large, very complex machine.

I found one particular feature of the book interesting. At the end was a section that listed the subscribers for the book. At the top of list were the names George Washington and Thomas Jefferson, who had Evans build new mills for them. To publish a book meant getting enough readers to commit in advance to buying it—a practice that certainly reminds us of crowd-funding today. An enterprising publisher or author would try to cover the costs before incurring them by pre-selling subscriptions to the book. Yet I was surprised to see the list of subscribers printed in the back of the book. I asked the docent at the mill about the list. He said it had the purpose of identifying others who bought the book and were interested in the same subject. If you were studying Evans's plans, you might want to correspond with someone else who knew the book. The docent pointed out that the subscriber list was a social network built into the book itself. A community of practice can develop around common projects, common tools, and common sources of knowledge.

The subject of books leads us back to Benjamin Franklin, who writes in *Autobiography* how he created a lending library. He recognized that it was difficult to obtain books in America and proposed to a group of friends in Philadelphia that each of them bring their books to a common room "where they could be a common benefit." Then he thought the benefit should be made available more widely, proposing a member-funded public subscription library. "So few were the readers at that time in Philadelphia, and the majority of us so poor," he wrote, "that I was not able, with great industry, to find more than fifty persons, mostly young tradesmen, willing to pay down for this purpose forty shillings each, and ten shillings per annum." He said that "on this little fund we began." In similar fashion and with little funding, makerspaces are started today to pool resources and provide a common benefit.

The 1970s was also a time when there was a flourishing of the do-it-yourself in books and magazines. The basic theme was taking control of your life, generally rethinking material culture and community.

John Seymour's *The Self-Sufficient Gardner* was a practical guide to producing your own food with rather ambitious goal of becoming self-sufficient. I have my forty-year-old copy. *Mother Earth News* magazine carried plans for building your own log cabin, yurt, or geodesic dome. *Our Bodies, Ourselves* was a pioneering user manual for women written by women. "We were just women; what authority did we have in matters of medicine and health?" the group of women in Boston wrote. They challenged the medical establishment by asserting the value of their own individual and collective experiences. They described the papers, which were originally typewritten and photocopied, as a tool that "stimulates discussion and action, which allows for new ideas and for change." They sought to provide information about anatomy and physiology that demystifies the human body and how it functions, initiating a "collective struggle for control over our bodies and our lives."[28] These books were designed to inform and shape our actions, another example of "what you can do with what you know."

The *Whole Earth Catalog* was first published in 1968 by Stewart Brand, who was an inexhaustible scout looking for new and old things to make part of an emerging counterculture. Eventually he was joined by contributors that included Kevin Kelly, Bruce Sterling, and Howard Rheingold. The catalog's motto was "access to tools." *The Whole Earth Catalog,* and its predecessors and successors, *CoEvolution Quarterly* and *Whole Earth Review,* were products for settlers of a new frontier in need of new tools. The *Whole Earth Catalog* audaciously stated its purpose:

> We are as gods and might as well get good at it. So far, remotely done power and glory—as via government, big business, formal education, church—has succeeded to the point where gross defects obscure actual gains. In response to this dilemma and to these gains a realm of intimate, personal power is developing—power of the individual to conduct his own education, find his own inspiration, shape his own environment, and share his adventure with whoever is interested. Tools that aid this process are sought and promoted by the *Whole Earth Catalog.*[29]

The *Whole Earth Catalog* represents a mind shift in defining progress in personal rather than industrial terms. I can't think of any other paragraph that so clearly represents ideals that are reflected today—optimistically so—by the Maker Movement. It is a new world that waits to be created, a future that can be lived in part today if we aspire to make it real.

As an aside, The *Whole Earth Catalog*, like Oliver Evans's book, lists its retaining subscribers on the same page as its "Purpose," naming those who contributed funds in advance so that the book could be published.

Kevin Kelly continued the tradition of the *Whole Earth Catalog* in his own way, self-publishing the massive "Cool Tools," a collection of tool reviews from the Cool Tools website. In a talk he gave at MakerCon in 2014, Kelly said that "revolutions begin with new tools," paraphrasing Freeman Dyson. "Every invention is a tool to make newer things," Kelly said. "When we make a new tool, we make a new way of seeing, which leads to new ways of knowing." Tools can lead to transformation of culture.

Kelly talked about how the 1848 gold rush brought people with a certain to California. It was a belief in the importance of doing things on your own and that you didn't need anyone's permission to do so. He sees that spirit carried through into the 1960s with the Free Speech Movement, coming out of Berkeley, California, and the Human Potential Movement, coming out of the Esalen Institute in Big Sur, California. Out of the culture of experimentation, which included music and drugs, came the *Whole Earth Catalog*, what Kelly called a "do-it-yourself bible" that was "exploring what we could do as individuals or as a small group." He said that the catalog had "alongside the tepees, the beekeeping equipment, and the macramé and the hand mills for grinding your own flour, the first listing of personal computers." All of this came from California, "where there was the least resistance to new ideas," he said. The Homebrew Computer Club also got started here with Steve Wozniak and Steve Jobs, in Berkeley and later in Menlo Park. Kelly said two strands came together: a revolution in tools that were developing along with a deep interest in human exploration. "This confluence came together in the personal computer," he said. That confluence "sparked some of what we are seeing again in the Maker Movement."[30]

In some ways, the personal computer is such a powerful tool—an all-in-one tool—that it has taken us more than a generation to absorb it fully and gain a perspective on what it does and what it can do. Or, more importantly, what we want to do with it. Now we are once again exploring new tools, recovering knowledge that was lost, and looking at how we can make our own lives better and contribute to the common benefit of all.

The maker , and its reemergence in our culture, has as much to do with why there is a Maker Movement as any technology or economic trend. It represents a cultural mind shift: the empowered individual over the institution, the open and self-organizing network over the rigidly organized corporate hierarchy, experimental play over busy work, agency over apathy, creative expression over soulless conformity—the joyful life of the maker and producer over the contented consumer. It is a that integrates what can be seen as separate: manual and mental, science and art, engineering and craft, risk and resilience, practical problem-solving and world-changing imagination. One might say about the maker what Marvin Minsky said in *The Society of Mind:* "Much of [the mind's] power seems to stem from just the messy ways its agents cross-connect."[31]

8

Making Is Learning

Quin Etnyre was eleven years old, having started middle school in a coastal city in central California, when his mom noticed that he was just not enjoying school anymore. He was a bright kid, but he was uninspired by his days at school. She took it upon herself to look for something that might interest him. She found *Make:* and bought him a copy, and then brought him to Maker Faire, where he wandered around checking out the wild range of exhibits, from zany to practical. At a workshop there, he learned to solder circuit boards.

Quin went back home after the Faire and taught himself to code. Then he moved on to Arduino. He started making things, and the next year he came back to Maker Faire as an exhibitor to show off what he'd been making. He built a website called Qtechknow, filled with how-to tutorials and project guides in addition to a catalog of his products. By the time he was twelve, he was following in the footsteps of Limor Fried and other professional makers.

Quin built a shield for Arduino called ArduSensor, housed in a second circuit board that can fit on top of an Arduino board and can be used to plug in other components. Essentially Quin made a system for easily adding individual sensors to Arduino. One of Quin's ArduSensors is a fart detector. "There were questions," Quin said, "so many questions. How bad does it smell? Do I want to run away? We can develop technology to answer these questions." Who else but a twelve-year-old boy would come up with such a thing?

Quin can be found teaching classes, like Introduction to Arduino, to adults, donning his MIT T-shirt with the idea that he might go there someday. He's even been on industry panels, the little guy talking about electronics alongside the adult presenters. Quin keeps returning to Maker Faire, where he has won ribbons for various projects and has been a featured speaker. In 2014 he came to Maker Faire Rome, where he tweeted that he got to see the pope and meet the mayor of Rome. When I last checked in with Quin, he had launched a Kickstarter to fund the development and marketing of a new product; he had already raised more than $30,000, and hoped to meet his stretch goal of $50,000. He's had a wonderful series of adventures as a young maker, but it all started when making sparked his curiosity and opened a new world for him.

One could consider Quin an exceptional child. He does have the good fortune of having parents who are able to provide him with resources and experiences like a visit to Maker Faire or to help him by filling customer orders on the kitchen table. But in terms of intelligence and ability, Quin is just like lots of other kids. He just managed to find his own passion and find a community to support him. There are many more young makers like Quin who have become well-known through Maker Faire, such as Super Awesome Sylvia Todd, Joey Hudy, Schulyer St. Leger, and Audrey Hale.

In 2014, as Quin was preparing to enter high school, the school board extended an invitation for him to come and talk to them about his products, his website, and his visit to the White House Maker Faire. After he was done, Jim Hogeboom, the superintendent, asked him, "Where did you learn to do all those things?" Quin replied, "Outside of school."

Hogeboom didn't like that answer. Neither did Quin. There was no opportunity to learn making in school, Quin told the board. He added that he'd like to see the school create a makerspace so that other kids could learn what he learned outside school. Quin suggested that Hogeboom call me, and he did. He wanted to know how to create a makerspace and how to prepare his teachers to help kids learn to make.

Engaging more kids in making has become a passion for me. There are plenty of kids who are never given the opportunity to make, and yet if they were, they might respond just like Quin. It could allow more young people to take control of technology in their lives while also taking control of their own learning.

I identify with kids who are bored and looking for something to do. I was one of them. When I was six, I was diagnosed with a disease of the bone marrow, osteomyelitis, which affected my right leg. I was hospitalized frequently, for weeks at a time. I found myself in a room with beige walls, a black-and-white TV, and a bedpan. Although my parents came to see me, I was alone a lot and immobile. I struggled with boredom. I didn't want to be bored, and I found out that it was within my control. I could play in my mind. My imagination could take me places my body wished it could go. I could find ways to displace boredom with almost anything. I loved books because they told me about the world, but I didn't see reading as an end in itself, but a means to an end. I was interested in learning how to do things. Over time, I realized that I was a good learner and I could learn on my own.

My mother was a teacher, and in college I was an English major with a minor in education. Yet I didn't become a teacher, in part, because what I learned from student teaching was that classroom teachers had little time to help students learn. Instead, the teacher had to be in charge, speaking in a loud voice to get the attention of students and trying all kinds of tricks to maintain control of a group of restless, even agitated, students. To be fair, my student teaching in the mid-1970s in Louisville, Kentucky, took place in schools affected by the desegregation court orders that bused minority students to white suburban schools and suburban kids to inner-city schools. It was motivated by the noblest of intentions, but it was done without creating the necessary support and infrastructure for students, teachers, or the community.

I certainly wasn't thinking of children or teens when I created *Make:* magazine. Its intended audience was adults, yet it became apparent early on that kids were already part of it. Maker Faire was getting younger every year, with more and more children coming. Kids were—and are—engaged by what they see and experience at Maker Faire in a way that they pretty much never experience at school.

I began wondering what happens to those kids on Monday, the day after Maker Faire is over. I thought of it as "my Monday problem." If kids were inspired by the weekend at Maker Faire to become makers themselves, where would that go on Monday? I was reasonably sure that students were not given the opportunity to learn making inside school, unless they were exceptionally lucky. How could I address that? How could I find others who wanted to change that? How could I work with museum and library leaders, afterschool providers, teachers, parents, and the kids themselves? How could I advocate for creating a context that allowed students to play, make, and learn in informal and formal education settings? I believed it would change how children learn, particularly in school, and it just might lead to transforming education so that it focused more on how children learn best.

INFORMAL LEARNING

In some communities, there are opportunities for the kind of playful hands-on learning we see at Maker Faire, in places outside schools or after school, at Children's Museums or summer camps.

The Exploratorium is a cherished museum of art, science, and human perception in San Francisco, created in 1969 by the visionary Frank Oppenheimer, who believed in direct, experiential interaction with physical phenomena as a path to deep understanding. Oppenheimer believed in tinkering. I have worked with Karen Wilkinson and Mike Petrich of the Exploratorium's Tinkering Studio on many programs to explore tinkering as learning. We developed a series of programs for families on Saturday mornings called Open Make. Each session had a theme, such as making musical instruments or making things from metal. Participants figure things out by trial and error and learn from what others are doing around them.

I remember watching a young boy standing at an old Exploratorium pinball exhibit that allowed him to create and modify his own pinball machine playfield. I was fascinated how he moved a set of wooden pieces of different shapes in place to alter the path of the ball. Each time, after shooting the ball, he reconfigured the blocks, changing them to create new obstacles and new paths. There were no instructions, no guides, and he was fully engaged, each time making it new and more challenging.

As Karen and Mike say in their beautifully illustrated book called *The Art of Tinkering:* "When you tinker, you're not following a step-by-step set of directions that leads to a tidy end result. Instead, you're questioning your assumptions about the way something works, and you're investigating it on your own terms."[1]

In Pittsburgh, the Children's Museum has created Makeshop in a room that once housed a Mister Rogers exhibit. Mister Rogers was from Pittsburgh, and Makeshop retains a framed sweater that he wore, which apparently his mother made for him. The Makeshop is a very well-designed workshop, distinguished by its simplicity and warmth, a combination of natural wood and stainless steel. The space has college students as guides who help facilitate making experiences with visitors. Jane Werner, the museum's executive director, told me that typically what she saw in Makeshop was a parent or grandparent who would sit down with a child, and they would start making together. What surprised her was how long they were engaged, sometimes taking a break for lunch and then coming back to pick up where they left off. Museum experiences are often short, but at Makeshop, families were occupied for long periods of time.

One of the student guides in Makeshop, Kevin Goodwin, did his thesis for an early childhood education program on tape as a material for children. "Tape as in the masking, scotch, duct tape; the stuff you use to connect, seal, decorate, bind, measure, hang, attach, etc.," he writes. "What are the skills involved in using tape, and what are the necessary motor skills that need to be developed before a child can properly use tape?" Kevin spent one month in Makeshop watching toddlers trying to use tape. Cutting tape with scissors and tearing tape are both difficult for toddlers to do, requiring dexterity, good hand-eye coordination, and fine-motor skills.

A question about making and tinkering is often raised: Yes, children enjoyed it, but what did they learn? We have a false dichotomy between learning and play, when in fact the two go hand in hand. "We may think we are helping to prepare our kids for the future when we organize all their time, when we ferry them from one adult-organized activity to another," writes Stuart Brown in his book *Play.* "We may be depriving them of access to an inner motivation for an activity that will later blossom into a motive force for life."[2]

Finland is often cited as having the best public education system in the world, and Finnish fifteen-year-olds rank at the top of international test scores. One of the reasons is the emphasis on play, which can be freeform as well as guided. "Finland requires its kindergarten teachers to offer playful learning opportunities—including both kinds of play—to every kindergartner on a regular basis,"[3] wrote Tim Walker in a story in *The Atlantic,* wonderfully titled "The Joyful, Illiterate Kindergartners of Finland." Walker is an American teacher who has taught in Finland and was struck that American and Finnish kindergartens were moving in opposite directions, one decreasing play while the other looking for ways to increase it. According to former Tufts professor David Elkind, American children have eight fewer hours of unstructured playtime after school each week than they did twenty-five years ago, on average.[4]

Making can give kids the "permission to play," a phrase we put on T-shirts for Maker Faire. Divine Bradley, founder of the Future Project in Newark, New Jersey, talks about transforming schools by changing the culture: "We made a culture where young people felt like they didn't need permission to be great, to do great things. It's not a young people problem, it's usually an adult problem." I hope parents understand that they are in a position to give or withhold that permission. Adults should give themselves the same permission, so that entire families join in making activities and projects.

Curt Gabrielson, author of *Tinkering: Kids Learn by Making Stuff,* sees tinkering as authentic and personal learning. He is asked the "what did they learn" question quite a lot by parents and school administrators. "Heck yes, they are learning something," writes Gabrielson in his book. "And it may be the most valuable thing they've learned all week … this hit-and-miss, trial-and-success, seat-of-the-pants approach."[5] He believes that "tinkering may raise all sorts of questions in their minds that inspire them to learn more about what they're tinkering with, and it may start them on a path to a satisfying career, not to mention good fun on their own time."[6]

Curt himself is a practitioner and helped create the Environmental Science Workshop in Watsonville, California. The leaders in informal learning often call themselves "practitioners," not teachers. It reflects a different view of the adult's role not as a subject-matter expert but as someone guiding children through a learning practice. Practitioners trust their own experiences in establishing programs that work for any child, regardless of age, gender, or socioeconomic background. They trust the successful experiences of children that they witness with their own eyes. Practitioners know what is working and what is not. Perhaps we should trust them more, because they are so focused on the children.

In fact, until recently there has been limited research into the educational value of making, in part because if you define educational value as performance on tests, you find yourself trying to provide further justification for the current educational system. There aren't good ways to measure engagement, for instance. I am sometimes surprised by what educational researchers actually measure.

At an educational conference intended to provide a context for making and learning in education, I recall that members of the audience kept saying: we get what making is, but can you measure its impact on a child's education? I got rather frustrated and said: "Of all the things you think you can measure, tell me what is really working well in education, because I don't see it. Is what you're measuring making the child's experience in education better?"

Recently, there has been some research that does support the role of making in learning. Lisa Brahms and Peter Wardrip, researchers at the University of Pittsburgh who have studied the learning and making at Makeshop, write about making as a set of learning practices, which they define as "more or less coordinated, patterned, and meaningful interactions of people at work."[7] They identified a set of distinct learning practices that they see:

1. Inquire—an openness and curiosity to possibilities.

2. Tinker—"purposeful play, risk-taking, testing" and engagement with tools, materials, and processes.

3. Seek and Share Resources—sharing knowledge and expertise with each other.

4. Hack and Repurpose—to reuse components and combine them in new ways.

5. Express Intent—discover one's own interests and personal identity.

6. Develop Fluency—gaining confidence in one's ability to learn and do things.

7. Simplify and Complexify—gaining an understanding of new ways to create things that have meaning.[8]

Our formal education system should adopt the practices of informal learning and bring them from outside school to inside school.

OUTSIDE IN

Traditional instruction in school—formal education—is based on a model of content delivery that involves a teacher lecturing and a student reading a textbook. The imperative to students is: "Study hard so you'll do well on the test." Driven by the noble goal of ensuring equity in the educational system, the No Child Left Behind reforms mandated that teachers all teach the same content, and students all take the same standardized test. Many didn't agree with this direction for education, but the legislation was a political mandate tied to funding. Who could argue against a goal to provide every child access to an equally good education?

The reality is that the federal and state programs were horribly designed, in ways that eliminated the autonomy of both educators and students. School reforms introduced a command-and-control, all-business approach to schools that made education worse for nearly everyone. The regime of accountability and standards failed to improve how children learn in school and how teachers teach. The social agenda did not have a viable educational model behind it, and most of all, it really failed to get educators behind it. The reforms created a climate of blame and distrust in education. Its moral imperative turned out to be an empty promise. It was going nowhere long before the federal Department of Education began easing its foot off the pedal in 2015.

Paul Heckman, an associate dean at the University of California, Davis, School of Education and a progressive educator, once encouraged me by saying the Maker Movement was creating the "counter-narrative," something that would become necessary when the dominant narrative of standardized testing had lost its power. The idea of getting making into schools at times has seemed an insurmountable task to me. There was little political support and very little money. Making was deemed insignificant, taking place outside of school often by poorly funded organizations and led by heroic individuals who were seldom recognized for their achievements.

At first, I presumed that educators themselves would be as resistant to change as the entrenched bureaucrats were. But I began to see the opposite was true. Most educators got into the profession because they, like me, had a passion for learning. Seeing a growing number of children falling behind, they were open and receptive to change. They recognize that students are natural experimenters who want to know how things work and why things are the way they are. As teachers, they faced the very real problem that their kids were bored, not engaged—and that meant they weren't learning.

There are some teachers who tell me they do project-based learning (PBL) that involves hands-on activities. Project-based learning can be aligned with making, but there's an important difference. If students are doing a hands-on activity at the direction of a teacher, often to support a curricular goal, it is not a maker project. If the things students make have no personal value to them, even though it is physical, it is not a maker project. If all the kids are making the same thing at the same time in the same way, that's just unfortunate. Making can be so much more than that.

A whole lot of teachers started coming to Maker Faire, looking for inspiration as well as for new projects they could bring back to their classes. In 2013, we gave away Maker Faire tickets free to educators if they filled out a survey that asked them why they wanted to come.

Nancy Gittleman, a preschool and elementary visual art teacher in San Francisco, said:

> What I love about Maker Faire is that it's everything important: creativity, science, art, electronics, math, engineering, all together. It makes sense: everything's connected. In my classes, I use the maker attitude: ignite creativity and imagination and connect projects to the world through inquiry. I teach art, but more than that, I assist the students to explore materials and to invent. My best projects are ones the students devise and expand on.

A high school teacher from Gilroy, California, shared:

> I really enjoyed Maker Faire because I got to see a lot of innovation and ideas that I could bring back to my classroom. It also helped me to start to look at developing ideas to make the curriculum more accessible for my students through shrinkable cell models, career opportunities, and adapters that could make cell phones into microscopes.

A school psychologist told us:

> Because of their emotional problems, and often learning problems, many of my students have faced years of academic failure. Making things helps them to feel competent, productive, and sometimes to express emotions. They can often feel part of a community of makers and can share what they have made with family. Making is a very valuable tool to help these adolescents feel better about themselves.

A special-ed teacher wrote: "My students struggle with *all* learning. We made little wooden birdhouses. They had to figure the perimeter of the birdhouse, measure each side, make templates to cut out their paper, and cut the template using as little paper as possible. They never got that it was math!"

Debbie, an elementary school teacher in Cupertino, California, wrote:

> Students in my classes are very strong academically. They show their deeper understanding through hands-on projects, not just through traditional means such as testing and experiments. Projects also give students practice working in collaborative ways toward a common goal. Making allows students to use their knowledge and skills to bring concepts to life.

I was not alone in thinking that more children should be able to have in-depth, hands-on learning experiences in school. The Maker Movement was starting to transform education from the bottom up, unlike top-down federal and state educational reforms. It was largely due to the efforts of teacher champions.

A RENAISSANCE IN THE MAKING

As more parents, teachers, principals, schools, and school districts recognize the role for making in education, we have seen a renaissance in exploring alternative ideas about learning that come not so much from current research but from old sources. The word *renaissance* literally means "rebirth," but it's more like a rediscovery. The Italian Renaissance was a rediscovery of Greek and Roman knowledge and wisdom that led to a flourishing of new ideas. Today we're reengaging with the prominent educational philosophies of the first half of the twentieth century. Maria Montessori and John Dewey published their philosophies of education around 1920, while the developmental psychologist Jean Piaget wrote about his findings in children's development in the 1950s and 1960s, and Seymour Papert explored the positive ways that technology could help children learn in the 1970s.

Born in Italy in 1870, Montessori developed methods of teaching and creating environments conducive to learning that at the time were considered revolutionary. Montessori used object-based exercises to develop an "education of the senses." Her exercises encouraged children to explore the sense of touch with their eyes closed. She set out metal containers of water heated at six degree intervals; blocks made of three different woods that differ in weight by six grams; and other blocks with alternating strips of smooth paper and sandpaper. Children were asked to recognize the differences and respond by placing the objects in some order.

Montessori believed that this education of the senses was important for the child's ongoing development as a learner. She wrote:

> To teach a child whose senses have been educated is quite a different thing from teaching one who has not had this help. Any object presented, any idea given, any invitation to observe, is greeted with interest, because the child is already sensitive to such tiny differences as those which occur between the forms of leaves, the colors of flowers, or the bodies of insects. Everything depends on being able to see and on taking an interest. It matters much more to have a prepared mind than to have a good teacher.[9]

Another hallmark of the Montessori method is to structure the environment and teaching so that children become independent, self-directed learners. She noticed that practical activities such as caring for the self or caring for surroundings, like washing, cooking, exercise, or gardening—hands-on activities that have a clear purpose—had an attraction for children and gave them a sense of accomplishment. Her methods emphasized choice, challenge, curiosity, and collaboration. She believed that children learn better when they can choose what to do and when. This philosophy of education later became known as "constructivist," meaning that a child must construct the world to understand it. Learning is active, not passive: we don't receive knowledge, we must construct it, as with building blocks.

John Dewey established "learn by doing" as an educational philosophy over a hundred years ago. He knew that we learn from real experiences, by interacting with the world and observing the results of our actions. Dewey wrote in *Democracy and Education*:

> To "learn from experience" is to make a backward and forward connection between what we do to things and what we enjoy or suffer from things in consequence. Under such conditions, doing becomes a trying; an experiment with the world to find out what it is like; the undergoing becomes instruction—discovery of the connection of things.[10]

We might call it "experimental play" as well as "experiential learning." Dewey emphasized the primacy of experience over the accumulation of knowledge. "I believe that education is a process of living and not a preparation for future living,"[11] he wrote. Dewey believed that kids were motivated to learn when they had some choice about what to study and some responsibility for deciding how. They were motivated when they worked on tasks that made sense to them, especially tasks that really needed to be done, and when they worked in the context of a shared endeavor. He believed that curriculum should be organized around real-life problems and questions and should involve important, current social themes. Dewey understood that education should prepare people for an unpredictable future with new challenges. He saw the need for us to become lifelong learners, experimenters, and problem-solvers, which is what distinguishes makers.

Learning by doing means we are getting hands-on, in a literal and figurative sense. I think sometimes we take "hands-on" too literally, as if "minds-on" were something different. The seal of MIT depicts a craftsman at an anvil and an intellectual in robes with the motto in Latin *Mens et Manus*, which means "mind and hand." As with that motto, "hands-on" should imply a personal way of understanding something through one's actions. The opposite of "hands-on" is "hands-off," which not only means not wanting to touch something, but also not wanting to engage with it or understand it.

The Swiss developmental psychologist Jean Piaget titled his book on education *To Understand Is to Invent*. It's a beautiful and unusual expression of a truth that learning is a creative act, a discovery of something new to the learner. He believed that an experimental process is at the heart of education, that children have to try things themselves to learn: "An experiment not carried out by the individual himself with all freedom of initiative is by definition not an experiment but mere drill without educational value."[12]

The practical challenge in applying Piaget's theory has been that it is inefficient. Engaging children in real-world experiments takes time, as they go through all the roundabout paths that are inherent in real learning. There are no shortcuts. While it may be more efficient to give them worksheets with questions that we know can be answered, we should not optimize for efficiency in education. It's not a very satisfying or rich learning experience. Cognitive psychologist Jerome Bruner wrote that authentic learning is "deep immersion in a consequential activity."[13] That is a great phrase to describe making. Immersion is not a quick dip, not a ten-minute exercise. How could schools be organized to offer more of these immersive experiences? The answer would be to focus less on subjects and more on the experience of learning. In other words, focusing less on teaching might open time for deeper learning.

Paul Heckman, along with his colleagues Robert Halpern and Reed Larsen, wrote a paper called "Realizing the Potential of Learning in Middle Adolescence." It highlights that students of this age learn best:

- when they focus in depth on a few things at a time
- when they see a clear purpose in learning activities
- when they have an active role: co-constructing, interpreting, applying, making sense of, and making connections
- when it is a shared activity within meaningful relationships
- when it allows for increasingly responsible participation within a tradition, a community, and a culture[14]

I might add to the list:

- when the work has personal value beyond school, whether aesthetic, practical, or civic.

In experiential learning, the teacher's role changes yet it is not less important. According to Piaget: "What is desired is that the teacher cease being a lecturer, satisfied with transmitting ready-made solutions; his role should be that of a mentor stimulating initiative and research."[15] Maker educators see themselves as facilitators, guiding but not taking control of the child's experience. "Facilitator" may sound like a neutral observer, a somewhat passive role, yet that is hardly the case. Instead of delivering content, the teacher has more time to interact directly with children, offering encouragement and support for their learning.

I like to use the analogy of coaching. First off, athletes (or the debate team, or chess club members) are doing something they love doing, so part of a coach's job is to build on that love and not crush it through harsh criticism, excessive competition, or humiliation. A coach organizes practice to improve how the players play individually and as a team. Practice is necessary for self-improvement and teamwork. The coach's role is pivotal in understanding the motivations of the players, keeping them motivated to stay focused in practice and to play to the best of their ability or beyond. Students who are listless and bored are not playing well. Students who are following orders and have no sense of agency or control will not play at the highest level.

The coach's role is not to play the game. A coach should be an astute observer of the players and give them feedback on their performance. An educator can be exactly this kind of valuable source of supportive feedback, helping students recognize their own challenges, encouraging them to persist and fight through frustration and self-doubt, and celebrating their accomplishments. The sign of a good coach is that the students need less and less coaching because they have internalized how best to practice and play.

In the 1980s, a professor of applied mathematics at MIT named Seymour Papert developed his own learning theories. Papert, who saw computer technology as a tool for children to begin teaching themselves, developed the idea of "constructionism." He argued that children will construct new knowledge the best when they are, in fact, constructing something real. Mitch Resnick and Jay Silver are continuing Papert's work at MIT, as described in chapter 7.

While they have been mostly ignored by mainstream education, the philosophies of Montessori, Dewey, Piaget, and Papert have flourished in informal learning settings like the Exploratorium and programs such as Reggio Emilia in Italy. It's analogous to Europe in the Dark Ages, when monasteries copied the books that preserved classical knowledge and held on to it until our culture was ready to rediscover it. Sylvia Libow Martinez and Gary Stager wrote a book for educators titled *Invent to Learn.* It provides more detail on educational research and practice. Their central thesis is that "children should engage in making because it is a powerful way to learn."[16]

SCHOOL MAKERSPACES

Most makerspaces were organized for adults, not children. I began to advocate, along with many others, that we create makerspaces for children in schools and libraries. Makerspaces could be understood as a new kind of library designed to serve the whole school and all subjects, not just one. In fact, school librarians are among those leading the way in organizing school makerspaces. A makerspace can organize the functions of an art studio, a shop class, a computer lab, and more.

I felt that a makerspace was a way to change the student's experience of school by creating a new place where learning could look and feel different. A makerspace is a workshop; it's not a place where you go to sit down. It's a place where you stand and move around to do your work. You might work at large workbenches with other people. You choose which tools to use and move to workstations for specific tasks. You see lots of creative work from other people.

My own particular interest was how makerspaces could attract students who were not performing well academically, and provide them with another path. I was less worried about the top third of students, and more interested in the others, the two-thirds who aren't doing well academically in school. I wanted those kids to have a chance to succeed as learners—and define what success means to them.

I decided that if I was going around the country talking about makerspaces in schools, first I should do something in my own back-yard, in Sebastopol, California. I approached the local superintendent, Keller McDonald, and asked him if I could support a maker class at the high school. As our *Make:* office was then within walking distance of the school, I offered to host it in our office, where we had some equipment and junior college students as interns. McDonald listened and said something that more people, not just educators, should say: "yes." He thought it was a good idea to try doing the class outside school. I also asked him if we could do something to get around block-scheduling, because kids needed more than fifty minutes. McDonald thought it could be done.

Several weeks later, McDonald called to say that he found a good teacher to lead the class. His name was Casey Shea, a math teacher who did woodworking at home in his garage. It's important to remember that for teachers as well as students, skills developed outside school can be brought into school. Another teacher, Dante De Paolo, taught biology and was a motorcycle fabricator on nights and weekends. He was excited to connect what he did outside school with what he did inside.

Casey would not be the last math teacher I'd meet who wanted to get involved in making. Many math teachers work with students who struggle in their subject, which often feels abstract and arbitrary. Casey became the champion for the maker class, and he did every-thing possible to pull it off, having to build it without much of an advance plan. The administration figured out how the class could run for seventy-five minutes. We agreed that the goal of the class was to build confident learners through the practices of making. Because the class was an elective, Casey had to recruit students. He set up tables at lunchtime and showed off projects. Many of the kids didn't know what making was. Yet thirty students signed up: mostly sophomores, and a few juniors and seniors.

Some were top students, but most were in the middle, and some were doing poorly. A senior who was an A student told me that the making class was the first time she got to do anything creative in school. Another rather quiet young man, whom I learned was struggling in school and living with his grandmother, chose to work on repairing and modifying a motorized scooter. Over time, he got better at explaining what he was doing. One day, Sanjay Gupta visited the class to do a segment for his program on CNN, and I was delighted to see this young man doing an interview on national TV and confidently explaining his project, even under the pressure of having a camera pointed at him.

Other teachers came to visit, many from outside the area. We also gave tours to administrators. These visitors witnessed students working without direction, their work indistinguishable from play. The room had a nice sound, like the buzzing of bees in the hive, as the students talked intently, sharing their work and watching what others were doing. If you asked them what they were doing, they would tell you about the specific problem at that moment.

I expected from administrators a series of skeptical questions about how making supported learning, and I anticipated arguments that it didn't align with standards. Yet when the administrators walked around the maker-space, they got it immediately. I remember one of them saying: "I have kids who need this." They saw how students benefit from a makerspace. For some of them, they could see that it would keep their kids in school—kids who were bored in school and kids who didn't respond well to academic programs. The administrators' questions were about how, not why.

In the second year of the program, the old shop class at the high school became available. It was a rough space with good light. What leftover shop equipment was there had been pushed into the back of the room. A portion of the room had been turned into a conventional schoolroom for an agriculture class. There were several stacks of solid blue ag textbooks on a shelf and a white board, a fitting symbol of uninspired instruction.

Saul Griffith, a master maker with an MIT PhD, and I applied for and won a DARPA grant to develop a playbook on Makerspaces and work with a network of schools to develop makerspaces as a pilot program. (Some people in the maker community were upset at me for taking grant funding from DARPA, the Defense Advanced Research Projects Agency.) I saw the goal as extending the Maker Movement into schools, moving what we saw happening outside in the communities into education. The role of the playbook was to make it easier for educators to understand what a makerspace can be and how to organize it. Educators don't need a large budget upfront to do so; they can start small and grow a makerspace organically. It could be built in a modular way, organizing areas for electronics and textiles, for instance, and sourcing tools and materials from the community. From my experience visiting lots of school labs and makerspaces, I felt that the biggest challenges for schools were getting the right people involved and creating a sense of community around the space. It was a question of fostering a creative, collaborative culture where students feel inspired to make and where caring and knowledgeable mentors provide support.

I realized in working with our teachers in the program that there were three distinct kinds of projects and interactions: directed projects, guided projects, and open projects. In any program of making, you'll find each type, but I think a program should be designed so that open making accounts for fifty percent of the work, while guided projects should be thirty percent and directed projects no more than twenty percent.

Directed projects should really be limited to teaching a specific skill or conducting safety training. Somewhat like traditional classroom education, the teacher speaks and the students listen; the teacher leads and the students follow. If you want to learn to use a drill press, you need someone to show you how to do it, and then you take your turn.

Guided projects are when a student follows an existing project, probably relying on a set of instructions. If you haven't built a potato cannon or a soda-bottle rocket, you should do so: there are lots of project instructions online. Neither of these are original ideas for projects, yet you can learn a lot by doing these projects. Interpreting instructions and working methodically to complete a project is harder than it looks.

The third type is "open projects," which is perhaps nirvana for makers who want to work on a project of their own. An open project is open-ended from start to finish. The student or team of students generate an idea for the project, and then figure out how to make it, doing research and learning new skills along the way. It might be prompted by a theme or challenge. An open project really is a product of students' creative design and technical skill sets. Some might see a problem in open projects in that children have brilliant ideas that they lack the skills to make real, but it's a good problem to let imagination get way ahead of reality. Projects may have constraints based on resources, tools, skills, or funding.

So students won't be able to build everything, but they can *imagine* everything. Often we'll see that imaginative ideas can lead to prototypes built using cardboard and tape construction. What they are able to build might barely resemble the idea, but it captures their intention beautifully so that it can be shared with others.

In my own town of Sebastopol we started at the high-school level. What has surprised me is how much interest there is for makerspaces at the elementary-school level. Melissa Becker, the principal of Meadow Elementary School in Petaluma, California, started a maker lab for her students. She had space but no money to buy materials. She sent a letter home to parents asking them to collect anything around the house that kids could use as materials for making, emphasizing the value of recycled things. She told them to put the materials in a zip-lock bag and send them to school with their children. Many did, and she got materials like used tubing, corks, scrap wood, PVC pipes, and more. She wrote to me, "I was in awe after my first half hour with each class: students were all engaged in building cool things and exploring how things work. Now I go in once a month for thirty minutes with each class, and we have a ball." She asked parents to volunteer as docents, and sixteen people responded, the best response she ever had. If you don't ask …

Parents and the community at large can play an important role as stakeholders in school makerspaces. Successful schools like Becker's reach out and engage them. They also send their kids out into the community to show off the things they made. Becker's students, for instance, made holiday ornaments and sold them to parents to raise funds for their maker lab.

Months later, Becker wrote me to say, "We want to move our lab forward and try some more maker-type activities that will involve circuits, soldering, and building." Becker applied and won a $13,000 grant from a local education foundation. Her efforts inspired other schools to get started as well.

Not every school can access the same grants, parental support, or community resources. In Cleveland, the Design Lab Early College high school serves 220 at-risk youth. I got a tour from Eric Juli, the school principal, and Sean Wheeler, an enthusiastic maker educator who has been developing maker programs for several years in schools in the Cleveland area. Both are truly committed to maker education and bringing it to the lives of the students who perhaps need it the most. The facility itself, built in the 1960s, was designed like a panopticon prison, so that a second-floor central office could oversee all the classrooms and activity. In an inspired move, Juli and Wheeler decided that this central office would become school's makerspace. The school doesn't have much of a budget for the makerspace, so they are doing everything on the cheap. They were still setting up the space while I was there, but they had basic woodworking equipment and a single 3-D printer.

Wheeler learned that a local company was throwing away a lot of wooden pallets, and Juli decided they'd stack them in the cafeteria, right in the middle of school. Students used the pallets to build an outdoor stage for the local Ingenuity Festival, a project that required them to learn woodworking but also design and collaboration skills.

I met Zuri, a student who had spent so much time in the principal's office over the last year that he recognized that Juli needed a gate at the entrance to his office. Working with wood from the pallets, Zuri was building the gate. By all accounts, Zuri faces many challenges growing up as a young black man. He has never done well in school, and just getting him to show up regularly is a win. Zuri speaks reluctantly, cautiously, keeping his focus on his work. Wheeler took Zuri to a local woodworking shop called Soulcraft for a Saturday workshop, where he started working on a bedside table for his room. Zuri remarked how working with wood that wasn't from a pallet was much easier and more satisfying. Zuri's woodworking allows other students and the staff to see him as someone who can be successful. Another student, Carrionn, came in showing us a cup he had 3-D printed, bubbling with pride. He only learned about 3-D printing and already was bragging about what he could do.

"I want kids doing real world work that matters," said Juli. "I see making as a vehicle to develop the and skills for first-generation high school students to succeed." He explained that finishing high school really is life or death for these kids because there is really no livelihood for those who don't finish school. "I am betting that making gives kids ownership over their lives. They can become members of a community that is building, caring, contributing."

While many private schools, with access to funding, have embraced making after deciding that it is good for their students, we need to support the same thing happening in schools of the inner city and in rural communities.

The promise of putting makerspaces in schools is that educators and the community can come together to change how children learn in school. It can happen when educational leaders are willing to reevaluate how physical space, the structure of the school day, the peer relationships of children, and technology can be leveraged to change the learning experience of all children. It can transform the culture of the school so that it supports creative work and the development of talent. It's a learning environment aligned with how children learn, seeing them as naturally curious and naturally good learners. We can transform our community as well by cultivating a creative maker and becoming a learning community for all people, young and old.

THE SMART GRID FOR LEARNING

The future of education is community-based, not campus-based. The number of options for learning outside of formal school is growing at an exponential rate, and we might think of them as many educational services with many different providers. What if all those services found in informal learning had the same degree of visibility as formal learning? For that matter, what if we unbundled the services found in formal education and made them visible as well?

Perhaps the best way to change our whole system of education is to organize all the resources in and outside the system so that more people can use them. The future of education will depend on leveraging many more alternative sources of learning so we can help more people discover and use them more effectively. Just as we are exploring alternative sources of energy, we ought to be exploring how to organize alternative sources of learning. Sadly, our kids have been hardwired to large school systems:

the coal-burning power plants of education. So I've come up with an idea: the Smart Grid for Education. To allow more people to generate and use alternative sources of energy, we have to replace a system that hardwires homes to the local power plant. The grid allows various new sources power to be distributed to anyone who is on the grid.

We should want students to have more choices based on their interests, and we also need it done in a more equitable way. Students who live in affluent communities find a rich set of learning choices and have a huge advantage over those with the fewest choices. Think about finding a piano teacher or an after-school robotics program. The so-called achievement gap is intensified by a lack of access to resources for personal development that are found in affluent and middle-class communities but not in poor neighborhoods.

The Smart Grid is basically a network of existing services or resources. As I talk about it here, it's just a concept; there's not even a prototype. It's a way of thinking how the Internet might be used to do what it does so well: create self-organizing, distributed social networks. Education needs to be as open as the Internet, as open as the Web, as open as open source. Instead of a cafeteria's set menu, it could be more like Yelp: to be used, monitored, evaluated, and reviewed. It should be as easy to find learning resources as it is to find a pizza place or a beauty salon.

If we had a Smart Grid for Education, we could see more educational offerings in a community or across communities. This would raise the visibility and value of informal learning, whether provided by individuals or institutions: classes, DIY workshops, lectures, projects, activities, community service, work-study, and other learning opportunities. Science museums, 4-H groups, music teachers, FIRST Robotics, Boy Scouts, and Girl Scouts could connect their current offerings into the grid. Even work-study programs, apprenticeships, and community service programs could be connected. Connecting a student to the grid could help them tap into resources that they or their parents didn't know existed.

A Smart Grid for Education could help students and educators find out perhaps not only what resources are in their community but also what is missing. We could use a map view to show where resources are concentrated and other areas where they are lacking, where student needs are not being met by informal or formal educational services. It would help identify opportunities to deliver new services in different areas.

The Smart Grid for Education could create a dynamic metering of a local learning community where activity becomes more visible as it is shared via Twitter and Facebook. The more visible it becomes, the more it will drive participation. How many hours do people living in Seattle spend on learning? How does that compare to Austin or Atlanta? The Smart Grid for Education would support a means of monitoring and visualizing the demand for educational resources by the community and their actual usage. It would help us create a more robust educational ecosystem in our cities and towns.

THE MAKER PORTFOLIO

As I've said, many stakeholders in education have asked about the research that shows what students learn when they are making, or how we might assess what students are learning. The questions bothered me, or perhaps the fact that I didn't have an answer to satisfy them bothered me. I knew that it wasn't about improving test scores. Then one day, in the shower, I came up with something that has become a mantra for me: *Making creates evidence of learning.*

Making is its own form of assessment for authentic learning. The products of making demonstrate what a person can do, and what they know. The process itself and the steps along the way can be observed, reported on, reflected upon, and documented in various ways and in various formats. It's show-and-tell, an ages-old form of demonstration. All of this is evidence of a person's creative contribution and their technical capabilities. In other words, what makers do and share becomes valuable evidence, which can be reviewed by peers, mentors, or independent experts.

A maker portfolio is a collection of projects, highlighting show-and-tell as well as providing links to notes, media, and documentation. This portfolio should live online and become something that anyone can discover to learn about who you are, what you are most passionate about, and what you've learned to do. Already students are creating this evidence online and sharing URLs that link to their work on Instructables, YouTube, and social media. These individual portfolio items are not yet well-organized; however, they are accessible online.

Before the Internet, the sharing of demonstrations, performances, and exhibitions was nearly impossible because you had to be there to see it. They were considered ephemeral. Today student presentations can easily and cheaply be captured and made available on the Internet. Now we can do better than using tests and grades to evaluate a student's ability and talent: we can present authentic evidence by sharing a URL.

Grades and test scores are a poor proxy for the talents of young people. We can develop better tools than transcripts. There are a number of existing portfolio software systems and websites, but often they are licensed to a school that thinks of the portfolios as their property. That's why we need an open-source approach to portfolios that gives students complete access and control to what is in their portfolio.

An open portfolio would organize all the student's projects, performances, and demonstrations that reflect authentic learning experiences, whether they occurred in formal or informal settings. The open portfolio is an alternative educational record for a student, much as a transcript serves as a record of a student's participation in a formal education system. But with the record open, there's never a need to pay for access or copies. (I've never understood why schools assume they own students' records, just like hospitals and doctors once assumed they own patient records.) The open portfolio is also valuable for the students to see their own learning progression and perhaps share with others, including their peers who might wish to follow a similar path. In its simplest form, the open portfolio manages links to online resources (photos, videos, blogs) online that reflect the student's accomplishments.

There's an important role for educators in any open portfolio system. Educators can authenticate participation and provide an independent assessment of student work. There could be awards and recognition of students based on the quality of the work in the portfolio, and as importantly, the progression that they show in the development of that work over several years. What if, instead of listing classes, a school's website was a collection of student portfolios showing the work of everyone in the school?

When it's time to apply for college, what if portfolios were accepted as part of the admissions process? Some colleges already do. If a student applies to an art or design school, they will be asked to present a portfolio of their work, most likely a physical portfolio. It's interesting that while a design school might look at grades as a basic sign of competency, it has a process to review the portfolios of prospective students to evaluate their talents and whether there is a good fit with their program. That doesn't happen in science and engineering schools, unfortunately. Grades and test scores are all that matters, and that's increasingly a problem. A uniform selection criteria ensures a lack of diversity by using a narrow view of prospective students. It might also contribute to the high dropout rates by first- and second-year students in engineering programs.

What I've learned from talking to engineering school professors and deans is that they want a more diverse student body, not just in the terms of gender and race but also in terms of talents. However, because of the admissions criteria, a more creative student is more likely to choose a less structured design program over a rigid engineering program. Another issue affecting the dropout rate is that some students are directed into engineering programs because of their math scores. Yet when they get in the program, they realize that engineering as a subject and its course-load don't appeal to them. A student with a maker might get into engineering school and then realize that the classes are all lectures and no lab. Their love of making things and lack of support for doing that in school can make them change majors as well.

MIT, as one might expect, is actually leading the way among technical universities, announcing in 2013 that maker portfolios could be submitted as part of the admissions process. Students would still be required to submit grades and test scores, but now they could actually show their own work and have it evaluated. Dawn Wendell, then assistant director of admissions at MIT, made an announcement at World Maker Faire, saying: "As we see students getting more involved in the Maker Movement, we wanted to give them a more formalized opportunity to tell us about that part of their life and why it's important to them."

Wendell said that MIT wants to attract students that are "already solving problems and building, playing, and creating, engaging in projects that they love doing." She added that "not all successful students at MIT are makers, but MIT is a welcoming place for makers, or students who want to become makers."

In my talks with the admissions office and with professors, I learned that MIT initiated the portfolio process because they wanted to add ways to find students who were interested in building things. One mechanical engineering professor said to me that he teaches a machine design class, and the fact that so many students have little or no experience with machines is a problem. He wants more students who have some working knowledge of machines, the way that the kids coming from farms once did.

Several years in, MIT is getting over a thousand maker portfolios submitted each year. Disappointingly, fewer young women and minorities are submitting portfolios than young men. This reflects the predominantly white male demographics of the Maker Movement today, but I hope that can be changed as making spreads throughout communities and schools.

The student's open portfolio has value outside education as well. In the workplace, people list credentials and titles on résumés, but they are not necessarily a reliable indication of what a person can actually do. What we've seen in the technology field is that a person's work is often visible online: they've designed web pages or apps, written a blog, taken great photos, created videos, contributed meaningfully in discussion groups, or developed code in open-source projects. People looking for talent come upon this work and follow the connection back to its creator. A business that needs a certain kind of programmer might examine GitHub to find an expert. Chris Anderson of 3D Robotics found Jordi Muñoz to be an expert on building drones because he was so active on Anderson's site, DIY Drones. One might look on Thingiverse to identify the people with really good 3-D design skills. Sharing work openly online can lead to unexpected opportunities.

The Maker Education Initiative, the educational nonprofit that I cofounded, has a working group on open portfolios, and they will be producing a report with their recommendations on how educators and students can use such portfolios, incorporating the best practices of those who are already doing it. It doesn't matter as much whether there's a particular system or software to do it. The more important thing to realize is that young makers (as well as the rest of us) should be encouraged to tell the story of what they made, why they made it, and how it works. They should share the work openly. It's part of cultivating a maker culture.

As I've mentioned, some young makers are already figuring out how to share their stories and find support for what they want to do. Audrey Hale was a fifth grader in Biloxi, Mississippi, who was looking for an idea for a science project. Her mom was a subscriber to *Make:*, and so Audrey went looking through back issues. In volume 24 she found a balloon-cam project that had a rig consisting of a pill container, a CD-ROM, a microcontroller, and a camera, along with three large balloons that would take it airborne. She thought of it as a satellite, and decided it would be perfect as a science project. Excitedly, she went to her mother and told her she needed some money to buy the supplies for her project. Her mother reminded her that they were a military family, and they didn't have money in the budget to buy her supplies.

Audrey accepted the answer and then went away for a while before coming back to her mother with a new idea. She proposed that she run a Kickstarter to raise the money she needed. She asked for her mother's help because she was under the age limit on Kickstarter for posting a project. Her mother said yes, but with healthy skepticism, writing to me:

> I actually thought it would be a good way for her to learn that sometimes even when you want something really bad and you ask for help, the answer would still be "no" or "not yet." Turns out that my passionate little scientist ended up teaching me the lesson that sometimes when you find your tribe, your passionate ideas become real.

Audrey's Kickstarter, which she called the Little Ariel Satellite, was funded with thirty-two backers providing $420 for her science project. She promised her backers a photo book of her project, providing tangible evidence that she did it.

Audrey built her satellite and entered her school science fair, where she won first place. At regionals, she won first place for engineering in grades 4 to 6, and then won the first place special award given by the American Institute of Aeronautics and Astronautics. Audrey and her family were also invited as participants in the National Maker Faire in 2015 in Washington, D.C. She met members of Congress, attended a reception at the White House, and got to meet other young girls and women who were makers, scientists, and technologists, including Megan Smith, the Chief Technology Officer for the U.S.

MAKING DISTRICTS

Albemarle County in Virginia is the home of Thomas Jefferson's Monticello and the University of Virginia. Pam Moran is superintendent of Albemarle County Public Schools, a large geographical area with twenty-six schools. She is a passionate educator, a person I immediately liked because she's fun to be around. She cares so much and loves to tell stories about her schools, her teachers, and her students. She still remembers our first conversation in 2010 when I expressed my worry than making would never find its way into schools because of the dominant focus on standardized testing. If I was once pessimistic, seeing what is happening in Albemarle County is grounds for feeling optimistic about change.

In most schools that I know, making was introduced incrementally, sometimes with the acknowledgment of school administration and sometimes not. One teacher from St. Louis told me that his makerspace started as a closet one year, then he got a small room the next year, followed by a bigger room the following year. Now he has a budget of $80,000 and he knows of twenty other school makerspaces in the city. What makes Pam and her district unique is that they haven't done just one thing to support maker education; they have tried to do everything across the entire school system over a period of years to cultivate a creative maker culture.

"I describe us as a maker district," she said. Every classroom is a makerspace, a lab, a studio. The high school library is a makerspace. On a tour of elementary and middle schools, I saw kids working in small groups, and the focus was on their work, not the teacher.

Pam told me, "We embarked on this work many years ago, and didn't know then that we would call it 'making.' When the No Child Left Behind law came out, we saw it was going to suck the passion out of the classroom, out of our teachers, and out of our kids. We said we have to do something about that." She said that the prevalent educational model of the twentieth century, carried over to the twenty-first century, was about subtraction. Schools removed art, field trips, and shortened recess. The worse kids do on standardized tests, the more that's taken away from them. They are required to spend twice as much time in a subject they fail at, and it means they have no time for electives such as art.

She put together a team to do what she called the "transformational work" of changing the school experience such that "students pursue their interests, to be able to learn through their hands as well as their minds, to wrap their heads around big ideas, and to do all the kinds of things that sustain us for a lifetime. We want to prepare them for lifelong learning." Pam wanted to make sure they were capturing at-risk kids and getting them to come back to school, stay in school, and do well.

Pam and her team had to look outside education to find what they were looking for. "We could not find schools doing what we wanted," she said. Her team visited the Tinkering Lab at the Chicago Children's Museum. They went to World Maker Faire. "I would send them videos from the Bay Area Maker Faire," she said, hoping that something would grow organically and spread throughout her district, what she called a rhizomatic process, referring to way a rhizome sends out roots and shoots to spread.

The team came up with a maker-infused curriculum that "changes how kids learn and who gets to learn." She wanted both the kids and the teachers to see making as part and parcel of everything they do at school. They replaced summer school with Maker Camp, taking a different tack for kids who had fallen behind in the school year. They saw kids who struggled in school have success at Maker Camp. Pam visited the kids at camp and saw one group who had visited the local animal shelter on a field trip and came back the next day wanting to make things for the animals. "They were rolling out dough for dog biscuits and cutting them out like cookies," she told me. "Do you know what shape of biscuits were? Mailmen."

Pam said, "We don't think of the maker work as a *program;* we think of it as a way of learning." She talks about underserved kids who are often excluded or left out of the opportunities for "rich learning work that's reserved for some kids but not those kids." As the schools have added more choice in ways of learning, it has made a difference: having choices motivates children to learn. "We have a dropout rate that's down to 2.3 percent, and for African Americans, it is even lower."

I asked her for advice for other schools. "I've got it down now," she said with enthusiasm. Like me, she gives a lot of talks to schools and school administrators:

> I call it the YELP model: The *Y* stands for yes. People come to you with ideas—teachers, kids, principals, parents. You have to be receptive in order to keep people feeling connected to their motivation. Someone came and asked me early on about having a makerspace in their school. I said I think we could do it. The *E* is engage. Who do we need to bring together? Who can we work with? Who is there in the community? The *L* is leverage. What resources do we have that we can leverage? If you lack resources, ask parents to donate tools and supplies. The *P* is prototype. What experiments can we try to get more kids coming back to school because they love school and get excited about learning?

Pam said it didn't take six months before she and her team knew that something had really changed. She is particularly proud of a student who came to their district from Brooklyn, New York. The kid had gotten into some difficulty and was sent to live with his grandmother. "We were scared that he was not going to graduate because he had basically nothing on his transcript." One morning, the student wandered into the library and noticed the music construction studio that had been built a couple of years earlier. She said,

This kid is a musician, but nobody knew that about him. He starts working on music in the studio. He's a rapper. All of a sudden, here's a kid who is coming to school at 7 a.m. to get access to the studio, and then staying all day until we have to kick him out. We had thought this kid would be a dropout by the winter holidays. But he kept working, doing work online and in class to grab the credits he needed. Now he still had to pass the state's standardized test, which in Virginia goes by "SOL."

She laughs, acknowledging that the acronym of the standardized test in Virginia could stand for "shit out of luck." She continued, "I got a call two weeks before graduation from one of my principals. She had this kid who wanted to talk to me, and he said proudly that he passed his earth science test, a ninth-grade course, which was a requirement for him to graduate. That kid walked at graduation. That's what choice does."

MAKING PROGRESS

In the fall of 2015, before World Maker Faire, I organized an education forum at the New York Hall of Science. As I set out to organize a series of five panels for the forum, I had the insight that something had profoundly changed in education. No longer were there a few of us on the outside talking about how making belonged inside schools; those who worked in education were now doing the talking about making in schools and communities. Carmen Fariña, chancellor of the New York City Department of Education, came to welcome everyone. She was happy to see that a third of the audience were teachers from New York City. She said that she believes in the kind of bottom-up approach to change represented by the Maker Movement. She added that her grandson was particularly excited to be going to Maker Faire.

At education conferences, I often begin my talk by asking the educators if they see themselves as makers. If teachers see themselves as makers, so will their students. I ask the question sometimes reluctantly, fearing that no one will raise a hand. However, even in the worst cases, I get more than a few people raising their hands, maybe ten percent. If I ask how many people love to cook or sew or work in the garden or play a musical instrument, more people raise their hands. I remind them of the many different forms of making. If I don't get everyone, I just tell them to answer the question "Are you a maker?" with "yes," because it's just a better answer than "no." If making is learning, then teaching can be making as well, perhaps one of the most valuable creative acts in society, encouraging children in the development of their own talents. I'd like to imagine that maker educators set a new standard for education: one where kids are excited about learning. In January 2016 I spoke to educators in the remote Humboldt County, California, town of Eureka, and when I asked the question, the entire audience of teachers raised their hands. That's progress.

The Obama White House has been a big champion of the Maker Movement, holding the first White House Maker Faire in 2014, when President Obama declared: "If you can imagine it, you can do it, no matter what it is. We imagined things and then we did them. That's in our DNA, that who we are."

White House staff in the Office of Science and Technology Policy, under the direction of Thomas Kalil, have helped raise the visibility of making within federal agencies and at the state level, organizing workshops to bring educators together. "It doesn't seem to be challenging for a high school to raise $1 million for a new football field, so it shouldn't be challenging for a school to raise $20,000 or $30,000 for a makerspace," Kalil has said. He encouraged me to start an educational nonprofit, the Maker Education Initiative, called Maker Ed, and came up with its one-line mission statement: "Every Child a Maker."

Kalil and his staff have also helped promote diversity by reaching out to leaders in historically black colleges and universities (HBCUs). I spoke at an HBCU workshop, where I was impressed by the amount of maker-related programs and spaces that were already in place, as well as the interest among students. The support of the Obama administration has developed recognition for making and makers across the country as well as internationally.

The Agency by Design initiative at Project Zero, a research group at Harvard Graduate School of Education, published a white paper in January

2015 called "Maker-Centered Learning and the Development of Self." "Maker-centered learning" means that making in school should be centered on the student's growth and development rather than curricular goals.

The Project Zero researchers emphasize "the power of maker-centered learning to help students develop a sense of personal agency, a sense of self-efficacy, and a sense of community." They did site visits and interviews with maker educators:

> Many of our interviewees talk about how maker experiences help students learn to pursue their own passions and become self-directed learners, proactively seeking out knowledge and resources on their own. They describe how students learn to problem solve, to iterate, to take risks, to see failure as opportunity, and to make the most out of unexpected outcomes.... Perhaps most strikingly, they talk about how students come to see themselves as capable of affecting positive change in their own lives and in their communities.[17]

The white paper also introduced the term *maker empowerment* to describe not just cognitive development but also character development. It is very consistent with the maker , perhaps dressed up in more academic language. Maker empowerment "is a kind of disposition that students develop—a way of being in the world—that is characterized by understanding oneself as a person of resourcefulness who can muster the wherewithal to change things through making." Maker-centered learning can empower students to live productive lives, regardless of their future occupation. "Not all students who are exposed to maker education will go on to become scientists, technology specialists, mathematicians, engineers, or carpenters. But perhaps, through high-quality maker-centered learning experiences, they might all acquire a sense of maker empowerment."[18]

What's good about the report is how it affirms the value of what maker educators and practitioners have been doing. It can be cited by those who need to base their decisions on research, although academic research on learning rarely has been the driver for change in the classroom. Traditional education and even its reforms have followed a "push" model, as defined by John Seely Brown and John Hagel in their book *The Power of Pull.* Push leads to compliance. When I first heard a group of educators use the word *compliance,* I was taken aback. Was it compliance that dominated their thinking? If so, then conformity, not creativity, will be its main product.

Maker-centered learning is "pull." Students decide what they want to do. Making is being pulled into schools by educators because they see it as a very effective way to engage students in learning. They do it freely because of what it means to them and their students. This kind of change is being driven by parents, teachers and administrators—and students—who see themselves empowered to make change throughout the educational ecosystem. Research is not leading the way; nor is theory. Actions and practice are.

Tim Lord, an actor turned educator, is co–executive director of the DreamYard Project, a twenty-year-old community arts center in the Bronx, New York. It was started "by a group of young people who once wrote a play about a place where kids could go to dream."

When I visited in 2012, they had just started an afterschool program called Dream It Yourself to introduce students to how things are made and how to use maker tools. They had received a grant to set up a small makerspace. What made it special was their emphasis on the arts and personal expression. They looked at the functional side of making, but they provided an artistic context.

"If they wanted to build a lamp, we'd nudge them to make the lamp a sculpture that lights up" said Hillary Kolos, who is now director of digital learning. "Not everybody knows what 'maker' is. We try to explain it, but sometimes I describe it as design and technology." In the diverse population of the Bronx, Hillary and her team try to keep things culturally relevant by combining arts and social justice. They had to try and define what a maker is in terms that were relevant to the community. They had to ask questions about who makes things in their neighborhood and who they know who is a maker. "If we are going to show them a video of someone making something, I want that to be a person that looks like them," she said. "It can be hard to find those videos."

The DreamYard Project started by providing arts programs in schools along with afterschool programs for middle school students at their community art center. Now they offer programs around making in schools, and they have their own high school, which also offers maker classes. This is a pattern we are seeing around the country: successful making programs in informal learning settings are increasingly becoming the focus for new, innovative schools. Why? It works well for students, and they are asking for more.

In the community, Tim works with many organizations, including the large school district. He said that collaboration requires "an emphasis on shared values, and we have strong core values—empower, create, and connect." He attributes their success to a collective impact philosophy that connects the whole community to meeting the needs of young people. "When we look at the outcomes we want for our children," he said, "we have to believe that none of us has the silver bullet. We all have to be engaged and responsible for what we can do together." Transforming education is as much about fostering a creative learning culture in our community as it is about changing our school system.

9

Making Is Working

Nick Pinkston may be inventing the factory of the future. He's smart, independent-minded, and a risk-taker. He has worn his hair thick and long and has grown a beard since moving to San Francisco from Pittsburgh in 2011. He's not yet thirty. He comes across as oddly professorial because he's intellectually curious, yet he's also a maker and an entrepreneur. Nick could be the new face of manufacturing.

Nick said he has ADD, but he's not at all fidgety during a long sit-down for a conversation. He talks fast and changes direction in an instant. His business, Plethora, is located in a semi-industrial area of San Francisco near Pier 80 that some might call "sketchy." Inside, Plethora is mostly a very large concrete slab with masonry walls: a super-size garage with CNC machines used mainly for cutting metals and plastics.

Plethora is in the business of on-demand manufacturing, optimized for custom jobs: one-offs, specialized work, short runs. The Plethora factory makes pieces and parts, not finished products, but it may one day do the assembly required to produce a finished product. For now, Nick thinks that China is better at that.

Nick wants to make creating a custom part from a design as easy as the push of a button. If you are a product designer in San Francisco, you can sit at your computer and design a part; his software will provide feedback as to whether it can be made or not; and then you can press a button to send a design file over the internet to the Plethora factory, where it will be directed to a machine. After the part is machined, it will be sent by courier to you. You'd have your part in your hands possibly within hours. Nick believes that he can automate the order-taking, the validation and testing, the interface that controls the machines, and the setup for the job. If he's successful, his factory would be replicated in many places around the country or the world—and he would put many machinists, as well as the job shops where they work, out of business. Nick explained:

> There are thirty-five thousand job shops in the United States. Who knows how many overseas? You go to those places. You call them on the phone. You send them your files. That's how it happens. It's basically way slower and more expensive to do that. What we're trying to do is to bring an Amazon Prime–like experience, where you just hit a button and it shows up.

It's also like Uber. Why have your own car if you can get access to one when you need it conveniently and affordably? Why have your own machines or a machine shop when you can get access to one? Certainly, this has been part of Mark Hatch's vision of TechShop: You get access to any machine when you need it for a monthly membership. Nick is going one step further. You don't even need to go to TechShop or a fab lab to access local machines.

When the Plethora factory is scaled up, it will be full of sophisticated machinery, but it will employ almost no one. As Nick sees it, he's taking manufacturing to its logical conclusion by automating it. There aren't really robots in this factory; what happens is mostly in the software that controls the machines. To be clear, it's not necessarily the ability to send design files to a manufacturer that is the innovation here. That has been happening for quite a while. One can send a design file directly to a factory in China, if it is prepared to do the work. Nor is it automation per se. It's relatively easy to automate a process to create the same object over and over again. It's much harder to automate a process when the objects are different with each job. That's flexible manufacturing. Nick is eliminating the human intermediaries along the way, such as a contract engineer. That engineer's role is to evaluate a design, decide if it can be made, how best to make it, and provide a quote on how much it will cost.

A consistent problem with design software is that a person can design something in software that no machine can make. This means that a designer needs an engineer to look at the design and provide feedback on whether a machine can make the design. The challenge Nick ended up focusing on was first to analyze a design file and determine if it could be produced, for example 3-D printed; and second, to provide a quote for the work. Various 3-D printing service bureaus like Ponoko were doing this work by human inspection until they began using Nick's software. Getting a quote for a job required a phone call and a discussion, plus a day or so of research before the quote came in. Now Nick's software can prepare a quote on the spot, instantly, through the Internet.

Nick was the only child of two parents in the military. They moved around a lot. He went to a Montessori preschool, which he said was "the last school I liked. Every other school in my life, I hated." Montessori was hands-on. "You could do what you wanted. They had all these books. I remember just going through all the books and learning that stuff." He and his family always went to bookstores, and he recalls loving books that explained how things work.

Nick's family moved to the Pittsburgh area, where he struggled with high school. "That's when I started doing race cars and rockets," he said. "I used to hack that kind of stuff. Cars were a great thing. Everything in making can be done in a car: electrical, mechanical, chemical, everything. So, after sixteen, I was a car guy."

He went to the University of Pittsburgh and chose bioengineering because he was obsessed with nanotech, having read K. Eric Drexler's book *Engines of Creation* when he was younger. "No one in my class cared about making things," said Nick. "A bunch of them were good at math and thought they would be engineers. Even the professors really didn't care about it." He expected to be spending a lot of time in the lab, but he wasn't getting that opportunity. Instead, he started working with a race team for SAE International. "Because I could actually make stuff, they welcomed me on the team."

He decided to drop out of engineering. As much as he wanted to leave school, his parents wouldn't let him, so he came back as a business major, which gave him more time to do other things. He had read Neil Gershenfeld's book *Fab*, and he started reading *Make:*. He discovered that there were people like him, online but also in makerspaces, and he visited NYC Resistor, Hack DC, and Pumping Station One in Chicago. He thought Pittsburgh needed its own space. He eventually turned his shop into the first makerspace in the area, called HackPittsburgh, which he founded with Matt Stultz.

"HackPittsburgh let me know about other people who wanted manufacturing," and he would often connect people who needed manufacturing services with local manufacturers—job shops. "I would manually connect people and take a commission." Nick realized that perhaps he could create a way to make those connections online, and he began trying to automate the matchmaking between product designers and manufacturers.

However, there were a few problems. One was that running a business out of the makerspace created a rift. "People would think, 'Nick's a business guy. We're hackers.'" He eventually left and moved to San Francisco, which he believed would allow him to find more funding for his next start-up, which became Plethora.

I met Nick after he moved to San Francisco and he had organized what became an important monthly meeting: the Hardware Startup Meetup. Typically, it takes place in a warehouse in the Dogpatch neighborhood, where a hundred or more people gather with pizza and beer and listen to two-minute pitches from individuals who have a start-up or want to create one. They are wonderfully informal, rough-edged but friendly meetings that help makers who are looking for additional talent, funding, or just feedback. Nick became known as the hardware guy. Nick's model of the Hardware Startup Meetup has spread to dozens of cities, creating connections and providing momentum to those who want to turn their idea into a hardware product or service.

He pitched the idea for Plethora to investors. "I probably pitched eighty investors to do a seed round. It's like throwing things against the wall. Most said no." He raised $1.5 million, half of that coming from Peter Thiel's Founders Fund. "They're into countervailing ideas. They liked my contrarian angle." I asked him what he learned from fundraising.

"I learned that I have to be the guy to learn it. If I am the problem, I have to fix myself. I had to keep trying and failing and just knowing that if you keep pushing, you'll do it." A year and a half in, Plethora was up to thirty employees. The irony is that Nick is now recruiting those engineers who were really good at math but didn't really make stuff.

Nick sees his initial audience as "engineers with credit cards: They're buying prototype parts, and the parts are used for a machine." Plethora had made microscopes, robots, coffee machines, other 3-D printers and mills, and factory components. I asked Nick if he sees Plethora as the future of American manufacturing. "I don't think of it as American. It's going to be everywhere; in China, too. We'll deploy this to Earth," he said, laughing like a mad scientist. Plethora could be a way that manufacturing, which has been centralized, is decentralized and once again distributed regionally.

His focus, though, is not geopolitical but on the individual:

> I think, what capabilities does a random person have? A maker should be able to say, "I want to make a new coffee machine." Right now, you'd need a million bucks to build one if you want to build one that's good enough for sale. The reason is that it's really expensive and time-consuming to get the factory set up to make it. Having that local is really nice. It gets you more iterations. We often think about it in terms of giving engineers superpowers. The whole point of Plethora is that one person can do the entire thing. This is a factory that will allow you to make anything. It's not just milling or any one process. It's about the entire process of engineering. At Plethora, we will not rest until all hardware is like using a magic wand.

That's his enthusiasm to transform the nature of manufacturing speaking, or his ADD.

A factory that allows makers to make useful things without needing the skills of a machinist, Plethora is a perfect fit for creators in the creative economy. It's not such good news for the people who work in machine shops. Factories were the backbone for industrial America. Pittsburgh was a steel town, as were many Midwestern cities such as Youngstown, Ohio, and Rockford, Illinois, that flourished because of their steel mills. Yet in the 1970s, many large factories began to decline and eventually were abandoned, creating the so-called Rust Belt, with few opportunities and high unemployment. Towns that had expanded because of a single industry were left devastated and remain so a generation later. Detroit's decline from one of the largest and most vibrant cities in America became the symbol for the disruptive changes impacting factory towns. Even the auto companies, while they continue to make cars, need fewer workers. Membership in United Auto Workers declined from 1.5 million in 1979 to about 391,000 in 2014.

I remember a visit to Winston-Salem, North Carolina, where a community leader told me that forty years ago a young person could drop out of high school, walk into a factory, and land a job and a middle class life. This person, likely a man, could afford to own a home and a car, and raise a family where the kids were likely to go to college. However, in Winston-Salem, the three main industries—textiles, tobacco, and furniture-making—have closed their factories. Today that young dropout doesn't have a lot of options, except a low-paying service job such as a clerk at Wal-Mart. He or she probably continues to live at home. It's not a path leading to a middle-class life. It's as though the path is no longer there to follow.

DETROIT'S LEGACY

In 2009 I had the idea of bringing a Maker Faire to Detroit. I wondered if Maker Faire, which tapped into the Bay Area's creative culture, could work in Detroit, which was at one of the lowest points in its history. I had no personal connection to Detroit, but I came to feel that Detroit mattered to me. I knew that the people of the region had making and manufacturing in their blood. I wondered if they could tap into that power to reinvent their city.

I found a partner in the Henry Ford Museum, which helps us understand both the agricultural world that Henry Ford left behind and the world he was part of reinventing with the automobile. In many ways, Ford was a lot like Steve Jobs. Ford didn't invent the car, but he understood before anyone else how much it would change our lives if each of us had a car. Similarly, Jobs didn't invent personal computers, but he understood what might happen if we each owned one. Jobs had a vision of what personal computers could do for "the rest of us."

I began making trips to Detroit to look for makers. One of the first I found was a young entrepreneur named Andrew Archer who was looking to meet makers. That was the keyword Andrew used in his help-wanted ad on Craigslist. He needed help completing a large robot he was developing for use in auto factories. He wanted to target hobbyists who were curious and willing to figure things out for themselves. Jeff Sturges saw the ad and responded. Like Andrew, Jeff had moved to Detroit within the last year.

Andrew was offering only $10 to $12 per hour, but Jeff thought it was more interesting than any IT job he might find. On the phone, Jeff told Andrew about a community meeting for Maker Faire Detroit at the Henry Ford Museum that Sunday, and they agreed to meet there. That's where I met them both in January 2010. Jeff had moved from New York City, where he'd been involved in the Sustainable South Bronx fab lab. In Detroit, he was able to buy a house for $500, and he rode his bike around town to live on the cheap. Jeff grew up in the Boston area but had a degree in architecture from Cranbrook Academy of Art in Bloomfield Hills, Michigan, so he knew Detroit. He returned hoping to create a hackerspace and develop school programs to get kids involved in making things.

Andrew had moved to Detroit from Duluth, Minnesota, where he'd started his company, Robotics Redefined. He was using off-the-shelf components to design new kinds of robots for factories. He said he had a dozen contractors working for him and had sales worldwide. I interrupted him to ask how old he was. "Twenty-one," he said. I smiled. I immediately thought that finding people like Andrew and Jeff was a good sign for Detroit, and that makers were already connecting with each other.

Jacqueline Campbell Archer is Andrew's mother and his financial officer. As a single mom raising Andrew, she recognized that he had unique gifts that amazed and baffled her. "As a kid," she said, "if he went to sharpen his pencil, he'd end up taking apart the pencil sharpener." From age six, he took over the garage, bringing home things from yard sales or dumpsters. "Andrew liked anything with a cord," said Jacqueline. Once he hauled home a toilet so he could see how it worked. He then turned it into a fish tank. She would buy him tools instead of toys as birthday presents. He built a capacitor from cookie sheets and mineral oil. For a ninth-grade science project, he built a two-foot-tall Tesla coil, something his teacher didn't believe he—or anyone his age—could do. He was about to demonstrate the Tesla coil in class, but the teacher was so afraid of electrocution that he made him shut it down. "I was really comfortable that I was a person making weird stuff," said Andrew. He was mostly bored in school, feeling held back from exploring what interested him. He didn't like sports. He didn't connect easily with his peers. He dreamed of building a private spacecraft in his garage that would take him away to a new world.

More practically, he noticed that the rich kids in town all had mopeds to get around on. Unable to buy one himself, he began hacking one together from four bikes and an industrial weed-whacker engine. Jacqueline worried about her son. When he was young, she had taken him to the Mayo Clinic to consult with specialists. She learned that Andrew had a genius-level IQ, but she could easily see him dropping out of school or getting involved in drugs. High school was not challenging enough for him, and she sought to enroll her fourteen-year-old in courses at Lake Superior College. To do that, she had to sue the local school board.

One positive experience for Andrew was his involvement in SkillsUSA and its annual competitions. In tenth grade, he entered the robotics competition and won third place in the state. "They gave us a robotic arm and a box full of components to build an automated assembly," recalled Andrew. "You had no knowledge going in of what you'd have to build." The next year he was the state champion, traveling to the national competition in Kansas. A year later he was the national champion, requiring just four of the eight hours allotted to complete his project. Andrew graduated from high school in 2006, and a year later completed his degree in robotics at community college. He'd already completed a degree in machining at sixteen. "I was planning on going to Carnegie Mellon in the fall of 2007, but I decided not to," said Andrew. While he liked the university, he didn't want them owning what he worked on as a student. "I decided instead to pursue my own thing," he said, and started the robotics company that year.

At high-school graduation, Andrew saw his biological father, Bryan Fisher, for only the second time. Fisher, an inventor, had developed industrial baling equipment and built a successful company, Excel Manufacturing. From the short time they spent together, Andrew thought that the two of them were identical. "What he was thinking, I was thinking. He'd approach problems the way I'd approach problems, and we'd come up with the same solutions and say it the exact same way," said Andrew. "It was very strange." Nonetheless, Fisher remained distant. Fisher "had his own set of issues and stayed away, thinking he would probably be more of a negative influence on Andrew," said Jacqueline.

Both Andrew and Jacqueline say that Fisher was consumed with his own success, living life in the fast lane. In April 2010, Fisher was found murdered in his home, part of a triple homicide by a tattoo parlor owner involved in selling drugs and running an escort service. Fisher's company website said about the forty-six-year-old founder: "All who knew Bryan knew he possessed an all-consuming passion for power and precision, which manifested itself through his love of airplanes, cigarette boats, Ducati motorcycles, and scary fast sports cars. That same passion guided his equipment designs and broke the industry mold."

Andrew visited Excel Manufacturing the week after his father's death to meet his employees and take them out for an informal dinner, something they said that his father would have done. "I tried to take away just the positive things," said Andrew.

Like his father, Andrew has a fascination with motorcycles. On a summer night in 2009, Andrew was riding a Ducati Hypermotard and hit a culvert. "When I crashed, my first thought was—*oh, God, my bike*." He didn't notice at first that he'd nearly torn his thumb off and his foot was crushed. He rode the motorcycle to the hospital. His injuries, which included a lacerated spleen, kept him in the ICU for several weeks. Doctors worked to reconstruct his foot, and he used a walker for the rest of the summer. His planned move to Detroit would have to wait until October.

Jacqueline helped Andrew find a place overlooking the Detroit River in a brick building near downtown. He set up a small workshop where he could work whenever he wanted. There's orange tape on the wooden floor for testing some of his line-following robots, and his furniture comes from antique stores. After Andrew connected with Jeff in January 2010, they began working together to meet a March deadline for the factory robot. "I give everyone a test to find out what they can do," said Andrew. "Jeff got ninety-four percent on the test. It's really hard. Electrical engineers coming out of school would get about sixty-four percent on the test. That Jeff did so well is really odd, because he has an architectural background."

Jeff, a young thirty-three, started out by assembling circuit boards, doing surface-mount assembly, and learning how to test the boards. With his excellent people skills, he soon began doing project management. Jeff also recruited Bilal Ghalib, a maker from Ann Arbor, Michigan, who had organized the All Hands Active hackerspace there. Bilal's job was to write the laser-scanner interface for the robot. "I just threw it at him," Andrew said. "I didn't give him any instructions, and he did it."

Andrew was satisfied that he and his band of hobbyists worked so well together. "The hobbyist way is a really effective way to do things," Andrew told me. "We're utilizing tools that are available to everyone." He wants Robotics Redefined to become a kind of think tank for building things. "I want to do some things that are unconventional," he said.

Meanwhile, Jeff was also looking to find a place in downtown Detroit to set up a hackerspace. On a cold March day, he was in the Eastern Market district looking at vacant buildings in the old meatpacking area. The buildings smelled of dried blood and worse; they were in terrible shape. Pieces of mail were strewn about the floors, a good many of them delinquent tax notices. Jeff could see only the possibilities for each space, believing that they could be transformed. He liked the support he was getting from the management of Eastern Market, the location of the city's largest farmers' market and an area in need of new occupants now that the butchers had left town.

By mid-April 2010, Jeff had formed the hackerspace OmniCorp Detroit with a group of makers including Bethany Shorb and Andrew Sliwinski. With a grant from the Kresge Foundation, he started developing an entrepreneurial community workshop to build tools for urban farming, in association with Earthworks, a leader in Detroit's urban agriculture movement. He opened the Mt. Elliott Makerspace in the basement of a church, offering soldering and electronics workshops for children on Sunday mornings. He chose to work with the church not because he was religious but rather because it already had a community. "This is what I wanted to be doing," Jeff said. "This is why I came here."

Detroit had a variety of talented young makers. It was like they came together, formed a band, and played their own music for a while before breaking up, with each going their own way. Jeff remains an activist who believes that making is a way to introduce new skills and confidence to people living in disadvantaged neighborhoods, many of whom are left behind in the new economy. His model of engaging children through faith-based communities should be propagated widely.

Bilal Ghalib is a man without a category. He left Detroit and headed for the Bay Area, working for Instructables for a while. He spent a lot of time visiting makerspaces in Egypt and Lebanon and connecting makers in the Middle East. I once had dinner in Dearborn with his parents, and Bilal wanted me to help explain what he was doing with his life. They just didn't seem to understand him, and he made it difficult for them: he didn't have a regular job. He may never. He is a free spirit, an extra-extroverted maker.

Bethany Shorb runs a small business, Cyberoptix Tie Lab, making custom screen-printed neckties that combine a punk and geek sensibility. I visited her shop in a funky building in downtown Detroit, where she had four employees who were processing orders or making the ties. She makes locally but sells globally online: "Nobody in this neighborhood can afford to buy what I make."

In 2010, Andrew Sliwinski had his own interactive product design firm in Detroit. In 2011, he moved to the Bay Area and became chief maker at DIY.org for several years. In 2015, he moved to Boston, where he is now at the MIT Media Lab in the Lifelong Kindergarten group, working in Mitchel Resnick's group on the Scratch team. Andrew Archer took over as CEO of his father's company, Excel Manufacturing, a Minnesota-based maker of industrial baling machines for the recycling industry.

Several strong makerspaces cropped up, getting their start right around the first Maker Faire Detroit. I3 Detroit was the first makerspace in the area, located in Ferndale. Nick Britsky is one of its ten founding members. He told me that I3 brought together people who were previously working alone in their garages: "We pooled our resources to buy bigger, cooler tools. At the same time, we have this giant brain trust. You can build better projects, because you don't have to be a master of all of those things."

Dale Grover organized a coworking garage in Ann Arbor. On a visit there, I met a group of ex-Ford engineers building an electric motorcycle. At Ford, they worked on one element of a car as part of a very large team. Now their small team had to do everything, from design to build. They were loving every minute of it. Dale Grover teamed up with Tom Root to launch Maker Works as a member-based fourteen-thousand-square-foot workshop in an industrial park in 2012.

When I went looking for a new future for Detroit, I found makers— or I went looking for makers and found a new future for Detroit. The small, independent efforts by people like Andrew, Jeff, Bethany, and Bilal: there were enough of them to matter, and increasingly they were connecting to each other. Along with many others, they represented a new creative and industrious spirit that was emerging in Detroit. It is still at a prototype stage that will require many iterations.

MANUFACTURING AT YOUR FINGERTIPS

Economic development wonks charged with revitalizing a place like Detroit tend to think big. They dream of attracting a large manufacturing company that would bring ten thousand jobs to the region and restore the old system. An example of this was an attempt in 2009 to establish Michigan as a manufacturing center for a new generation of batteries.

The Obama administration awarded $861 million in federal stimulus grants, and the state of Michigan chipped in $543 million in tax credits for four plants, the kind of large private-public partnership that offers good publicity for all involved. Except that five years later those plants had failed to thrive. Each plant had employed a few hundred people, and according to a 2014 story in the *Detroit Free Press,* they had failed to use up all the grants and credits made available to them. Big is not always better.

The Maker Movement has the potential to change cities like Detroit and industries like manufacturing because makers think small, and it empowers people to think for themselves and act independently. The Maker Movement is open where businesses are closed; it is a self-organizing ecosystem, not a hierarchy. It is inclusive, not exclusive. In this new ecosystem, anyone, large or small, can participate.

What makers need is low-volume, small-batch manufacturing. They need to be able to make from one to five thousand of a thing. Plethora is doing that for one factory, but what if there were online interfaces to all the factories and suppliers in a region? If they were aggregated usefully, a maker could identify available resources needed to get something made. That's how we find a restaurant or a hotel today. Literally, we'd have manufacturing at our fingertips. It would help link local, regional, and national manufacturers with all kinds of makers, small and large.

In the twentieth century, Jane Jacobs wrote about the development of cities, looking at what makes them vital for social, political, and economic life. In her book *The Economy of Cities,* she wrote that economic growth happens as new work is added to old work. This kind of innovation comes from individuals and small groups, working independently but also interdependently. "The little movements at the hubs turn the great wheels of economic life,"[1] writes Jacobs.

Small is not only an alternative to big manufacturing. It is what manufacturing has become in the United States. One of the National Association of Manufacturers' twenty facts about U.S. manufacturing is that "the vast majority of manufacturing firms in the United States are quite small." All but 3,626 firms of the 256,363 firms in the country's manufacturing sector have fewer than 500 employees. "In fact, three-quarters of these firms have fewer than twenty employees."[2] If we could connect more of them to makers, there would be more work and more ways to get things made in Detroit or across the United States. If you're a maker, nobody needs to know that your factory is actually your garage, your spare bedroom, or a community makerspace.

COLLABORATIVE PRODUCTION

The first factories in England and America were for textiles. In the mills of Lowell, Massachusetts, all the operations of textile production, including spinning, weaving, dyeing, and cutting, were organized in one mill. Previously they were distributed, and the work had to move from one place to the next. In Lowell, the entire process, from bales of cotton to finished fabric, was the product of the factory. The labor force was mostly young women, so-called "mill girls" who came from rural farming families and lived in boarding houses.

What might those mill girls think of fashion designer Danit Peleg? She is designing 3-D printed fabrics for use in her own clothing line. Danit was a student in the fashion design program at Shenkar College of Engineering and Design in Tel Aviv, Israel. She knew nothing of 3-D printing until she went to Burning Man and someone gave her a 3-D printed piece of jewelry. She was fascinated and began wondering if she could 3-D print a dress. "I was always curious about the connection between fashion and technology, but I never had the chance to work with 3-D printing before," she told me at World Maker Faire in 2015. She decided to explore the idea for her final school project, developing her own 3-D printed collection.

Before she could make a dress, she had to make the fabric, a bit analogous to taking cotton thread and weaving it to create fabric using a loom. "I realized that if I could create a flexible structure, a flexible material, it would act like a regular textile," she said. The regular plastic filament used in consumer 3-D printers was not flexible enough. She got a lot of help from 3-D printer enthusiasts in Tel Aviv. One person pointed her to a filament that was flexible. She experimented with making fabrics, some of which look like lace, others like chain mail. Once she had the fabrics, she began to design dresses. Everything her models wore at Maker Faire had been 3-D printed, even their shoes.

"I was really happy to join an incredible global community of makers who share their knowledge, designs, and time to help each other realize their dreams," she wrote on her website.[3] Danit envisions a future where 3-D fabric designs can be downloaded and dresses can once again be made at home. In reality, it takes a long time to print the fabric. Danit estimated that it took her two thousand hours to print the dresses that she had in her collection. One red jacket that she pointed out to me took about 350 hours. Danit believes that new thinner filaments will come that have properties more like cotton or polyester and will help improve this new method of dressmaking. She believes more people will once again be making clothes again at home, thanks to 3-D printers.

In London, at Somerset House, I visited a shop called Knyttan, where you could make your own knit sweater or scarf. In the middle of the store, inside a large glass and wood box, is a fully automated industrial knitting machine. There are iPads in the store where you can choose from a set of premade designs and then modify them by changing patterns or colors or adding personal touches. When you are satisfied, you can send the design off, and it queues for the knitting machine, where your garment will be made on the spot for you.

Looking at the knitting machine in the store, I thought it was a beautiful thing in a postindustrial era. It's like a wood-fired oven in a bakery. Production and consumption are united in the same place, connected and made visible again. Knyttan doesn't send your design off to a suburban warehouse or a factory in Asia. It creates the garment in the store, and it's fascinating to watch it being made. Like the Plethora factory, Knyttan is also an example of automated on-demand manufacturing. No person is knitting the sweater for you. Nobody is there even to take your order. Knyttan represents an example of distributed production, which can be localized. It's a textile factory disguised as a boutique store.

LOCALLY MADE

The local beer where I grew up in Louisville, Kentucky, was Falls City Beer, using the city's nickname for its location at the Falls of the Ohio River. The Falls City Brewery was incorporated in 1905, and its shareholders were local tavern owners and grocers. They wanted to establish an independent brewery as an alternative source for lager to the Central Consumers Company, a brewery that also owned taverns and controlled beer distribution throughout the city. According to the archives, "the brewer, acting as landlord and supplier at that time, caused the tavern keeper to be more or less at its mercy."[4] Falls City successfully broke the stranglehold of Central Consumers Company and became a popular local beer through the 1960s. In the 1970s, competition from national beer brands caused a significant decline in sales of local beer brands like Falls City, which was sold off in 1978. The triumph of mass production and mass media advertising made it impossible for local beer companies to compete, and soon all that was left was nostalgia for a local brand. National beer brands seemed to able to identify a beer that pleased pretty much everyone, and in truth, one beer was pretty indistinguishable from another.

In the 1980s, something happened to change that. A new generation of beer drinkers came to think of Budweiser and Miller as swill—uniformly without character. Some of the real beer lovers became home brewers. They could make different styles of beer at home. Increasingly, they could find beer-making supplies at a local store that also offered classes on home brewing. Some of these home brewers started sharing their brew with friends, and those friends began to encourage them to make more, and offered to buy it from them. Management consultant Peter Drucker wrote that "there is only one purpose in a business: to create a customer."[5] Well, sometimes the customers tell you to create a business.

Their businesses were microbreweries that produced distinct craft beers for a local market. Beer lovers were looking for a wider set of choices and were willing to pay for quality. Craft beer was better because it was fresh, just like food. The establishment of microbreweries brought back local beer production. Unlike national brands, they didn't worry about distribution because their beers were mostly consumed on the premises.

An article in *Forbes* by Erik Kain, "The Rise of Craft Beer in America," reported that before 1910 there were about 1,500 breweries, large and small, in the United States. That number fell to zero during Prohibition. After Prohibition, the number of breweries returned to about 750, half of what it had been at the peak. These larger breweries competed with each other, with the number of breweries steadily dropping until reaching a low in 1979 of under 50. In the 1980s, we began to see the number of craft breweries growing fast, while the number of large breweries remained constant.[6] According to the Brewers Association, in 2014 there were 1,871 microbreweries and 135 regional craft breweries in operation, the highest total since the 1880s. Craft breweries make up ninety-eight percent of all U.S.-operated breweries.[7]

James Fallows, author and longtime contributor to *The Atlantic,* said at MakerCon in 2015 that his own leading indicator for the health of small towns and cities in the United States was whether they had a craft brewery. The return of local brewing in this country is at least one piece of evidence that national, centralized production is not inevitable or a permanent fixture. Things change, and the monolithic nature of national producers can limit their ability to respond to change.

Many of the things that are made locally today are in the food and beverage category. Farmers' markets have provided an alternative retail outlet for locally grown fruits and vegetables, connecting producers to consumers. They provide a niche for local producers to thrive in. Yet local production doesn't displace national production as much as it supplements it.

There is a resurgence of interest in locally made products, particularly when a product reflects its origin. Kate Sofis started SFMade, running it out of TechShop in downtown San Francisco. More than a promotional vehicle for locally made products, its mission "is to build and support a vibrant manufacturing sector in San Francisco that sustains companies producing locally made products, encourages entrepreneurship and innovation, and creates employment opportunities for a diverse local workforce." In Portland, Oregon, there's Portland Made, a collective organized by Kelley Roy of ADX. As a space, ADX has a focus on artisanal makers who are seriously into a craft or trade and making something for a living. At Portland Maker Faire, I met David Lewis, who had started the Veteran Bicycle Company at ADX in a ten- by ten-foot space. A machinist and veteran, David wants to engage other veterans in building bicycles. ADX has been a place to start a maker business for many like David. When he needed more space for his business, he was able to share space with another ADX spinout, the Portland Razor Company, which produces handmade straight razors and strops.

Are there other categories of products where local production makes sense? What about furniture? My brother, Dan, owned several furniture stores selling brands like Thomasville and Bassett. For years he went to High Point, North Carolina, for its annual home furnishings trade show. Most of the furniture he bought was made in factories in North Carolina. However, in the 2000s, most of the production moved offshore, and in 2014 Thomasville, over one hundred years old, went out of business. My brother closed his stores.

"Furniture designs and designers are valuable. So are the people who make furniture," Josh Worley of London-based Opendesk told me. Opendesk is a platform for locally produced designer furniture. On their website, you can choose a furniture design from a collection developed by independent designers.[8] Then you can choose a fabricator, ideally someone local with a CNC machine, to make the finished piece for you. Opendesk provides a curated set of designs known to be makeable, and a network of fabricators that can make it. "We give the customer a fixed price that includes the designer's fee, the maker's fee, and a platform fee," said Josh. Designers get ten percent of the price, much higher than what they normally get for a product sold in a retail environment. The largest share goes to the person who does most of the work, the fabricator—about seventy percent of the price the consumer pays. "It's a three-sided marketplace," said Worley. Their concept is Open Making: an open, collaborative design-build network. Grounded in open source, this Open Making concept is also trying to respect and value intellectual property: a design is open-sourced, yet they include a fee for the designer as part of the total price.

Could a company like Opendesk become a competitor to Ikea, adding more choice and better quality, plus a sense that the consumer is more connected to the actual producer? Ikea is like fast food. Opendesk is more like a high-end sushi place where you can select your ingredients and watch the sushi chef cut and assemble your rolls.

Opendesk is collaborating in the United States on a program called MakeLocal with 100kGarages, a project that got started by Ted Hall of ShopBot, profiled in chapter 5. 100kGarages is a match-making service, pairing small businesses with CNC machines with customers who have work they need done locally. MakeLocal is "designed around the world" meets "made in your neighborhood." This kind of design-build collaborative production model could be extended beyond furniture, allowing for more personalization and customization.

The big challenge is to generate enough demand: "If we can get an Ikea customer and convert them, that's how it goes mainstream," said Josh. If more people adopt a maker , they will care more about where and how things are made. Hopefully they will prefer that which is better designed and better made.

Another example of collaborative production is 3D Hubs. The company, which got started in the Netherlands, is a network of makers who own 3-D printers and who would be willing to print anything for a fee. If you don't have your own 3-D printer, perhaps there is someone near you who does and is willing to print your job. You might even be able to walk over and pick it up when it is done. 3D Hubs is approaching thirty thousand service providers in their network. It's fascinating to explore the interactive map on 3DHubs.com and see if there are 3-D printers in your town and where they are located. Makers appreciate the service, using money they earn to help them support their own 3-D printing habit.

Services like Plethora, Opendesk, and 3D Hubs provide new alternatives for consumers, but they also provide a greater ability for all kinds of makers to work independently. Makers can start small businesses that provide local services around 3-D printers or CNC machines. There are also a number of jobs in engineering and industrial design that once required working for a corporation in order to have sufficient capital and access to a lab or other resources. That work can now be done independently, as freelancers or small agencies, using makerspaces. Engineers or industrial designers may find that they have more control over the development of their ideas. They could become entrepreneurs, bringing those new products to market through online marketplaces.

The manufacturing industry talks about a "skills gap" in the United States, which means there are good manufacturing jobs that go unfilled. In a few talks with the National Association of Manufacturers, I've learned that they believe that young people don't want to work in factories, and their parents hold the same view. The manufacturing industry would like to change that view, so they instituted Manufacturing Day as a nationwide effort to invite people to visit local factories to see what they do.

It's clear to me that they have more than an image problem. Factory work, at its best, might be seen as something you might have to do, but it's no one's choice for work you'd want to do. It's repetitive, manual work. It's impersonal. There is a cultural stigma around factories, the kind of images we retain from enormous steel mills with smokestacks.

My own tour of factory duty happened the summer after high school, when I worked in a Ford automobile plant in Louisville, Kentucky. I worked as a vacation replacement at a variety of jobs on the assembly line. I still remember the distinctive smell of the factory. We mostly worked ten-hour days.

I got initiated on a job during which I got in the car during the water test. I climbed in the driver's seat while the car was pulled forward on its chain into what was essentially a car wash. I had two jobs: to look for leaks and to check that the car manual was in the glove compartment. As I was about to get in one of the cars, another line worker made the car backfire. Everyone laughed at my frightened reaction. Someone tossed me a roll of toilet paper. It was a kind of welcome that never involved "Hi, how are you?"

Another job I had was testing the emergency brake. I stood in a pit as the car rolled overhead. I attached a caliper to the emergency brake line and tested the tension in the line. I did this procedure forty-two times each hour while carrying on a conversation with a coworker who was also in the pit and two other guys who stood above us putting on the back tires.

What I remember most is the number of guys who were my age, having just graduated from high school, showing up to start the rest of their lives as factory workers. They expected to be there until they retired. I saw their situation as hopeless, even though there was a certain camaraderie among them, always teasing each other and telling tall tales about girls, which somehow occupied them while they did these repetitive tasks. I couldn't wait to get a break from the utter boredom of the job and go outside to sit against the factory wall and read a book for twenty minutes. I could not imagine doing that work for the rest of my life, even though I met men who were happy to do it. They got to go fishing on weekends and vacations. I wanted to figure out what I wanted to do with my life, but I couldn't imagine working at a job in a factory.

I remember telling folks at the National Manufacturers Association that if they wanted to change the perception of factories, they had to make it personal. Manufacturing could be viewed as a creative process. What you can do with a 3-D printer is manufacture a 3-D object yourself. That's pretty cool. I told them that their message shouldn't be about jobs in manufacturing, but what people can do to become manufacturers. Those people were *makers,* and they looked at manufacturing in a different way. It actually wasn't about factories. It was about what you could build if you had access to the tools of production.

TALENT POOLS

The best companies and universities are considered to be good at developing talent. Yet they often fail at recognizing the talent that they have and misuse it. And there's a lot of talent outside those organizations. How does that kind of talent develop?

The word *talent* can make us think that it applies to a special group, leaving out everyone else. I would agree with designer Charles Eames, who said, "I don't believe in this 'gifted few' concept, just in people doing things they are really interested in doing. They have a way of getting good at whatever it is."[9] I see makers as trailblazers. They are among the first to go on this journey that will eventually be shared by nearly everyone.

How is talent discovered? If I were a record producer in the 1960s, I would be going to different cities and visiting small night clubs. If I were a curator, trying to look for new artists, I wouldn't be visiting museums, because I'd be looking for artists who had not yet been recognized by museums. If I were looking for people doing start-ups, I would not be visiting large companies. In all those cases, I would be looking for talent in the wild, talent that had not yet been identified or widely recognized. I'd be looking for emergent talent, not established talent.

Over ten years ago, I went looking for makers in garages, former industrial parks, dark basements, recycling centers, city dumps, salvage yards, old abandoned factories, and artist lofts. I had to go out in the community and search for talented makers. When we started Maker Faire, the maker community began to come to us, each of them having found their own way. It created a way for us to find talented makers and get to know them.

Many makers are found outside institutions like corporations and universities. They value the freedom and independence that they have by working on their own. A makerspace fosters collaboration, and it can provide a new way to develop talent as well as provide a place for people to discover that they have talents worth developing.

Micah Lande, an associate professor at Arizona State University, once used the phrase "additive innovation" in a talk about makers.[10] It plays off the term "additive manufacturing," which is a fancy description for 3-D printing. Additive innovation suggests that makers, working in the open and sharing their work, benefit from the existing body of work and add to it. Hal Varian, an economist at Google, calls it "combinatorial innovation."[11] Truthfully, that's how science technology has worked. Inventions—as well as creative works—don't come out of nowhere; they come out of the connections between people, ideas, and projects.

I have often wished there was an IMDb for maker projects, which is similar to the Maker Portfolio idea I discussed in chapter 8. If work can be defined as projects, knowing the roles of various contributors on a project becomes important, just as IMDb does for the creative and technical talent involved in movies and television. The best way to help talent develop is to make it more visible. In the creative economy, credentials and a career path will matter less than the expression of your talent in projects and performance. What will matter is actual evidence of real work, not a resume listing jobs held.

On a visit to Youngstown, Ohio, I was asked by a person who ran an incubator there how a place like Youngstown could become a hub for innovation. Clearly, they can't compete with a Silicon Valley model or even what's happening in large cities. My thought was to orient the incubator to attract recent college graduates—Youngstown State is the local university. Instead of these graduates going out and looking for a job, perhaps they could enter a program where they could develop a larger project. The program need not do much more than provide common space and promote interactions with others in the program. One might offer them a very small stipend—ramen money; after all, they are used to living like students.

The purpose of the program would be to encourage these students not to take the first job they can find, but instead to see if they can apply their talents in this unstructured way. They might fail in developing their project, but I'd hope they'd discover that they like this kind of work and the autonomy it offers. If so, you've helped create an entrepreneur or an artist, who still may need more years and experiences to develop, but who can choose their own adventure. If the person struggles to be productive in the incubator, they might decide that they are better off taking a job. Nonetheless, you have given this person the opportunity to explore a different path for success.

MAKE WORK FOR YOURSELF

A recent study by Oxford University calculated the likely impact of automation in the next twenty years in the United States leading to forty-seven percent of current jobs being replaced by automated systems.[12] I know that doesn't sound like a sign of prosperity to most people, but what if we are able to eliminate useless work and find more useful work for more people? We need more machines and even smarter robots that can do more for us.

Even if such machines do eliminate jobs, they also create new opportunities, or at least new problems to solve, as we've seen throughout history. The Luddites, of course, protested that automated looms would eliminate jobs for weavers. The Washington, D.C.–based Columbia Typographical Union in 1903 sought to go before Congress and "enter an earnest protest against the installation of typesetting machines in the government printing office."[13] Linotype typesetting machines automated the placement by hand of individual characters to form a line of type. At a linotype machine, operators typed in the text at a keyboard and the machine composed the line of type. The International Typographical Union (ITU), perhaps the oldest trade union in America, had 121,856 members in 1964. Eventually linotype machines were replaced by photo-typesetters, and typesetting became increasingly computerized. By 1985, the same year that the laser printer was patented, union membership was down to 40,000. By the end of 1986, the ITU ceased operations. Anyone at a desktop computer could do what typesetters once did. Those who learned to design for laser printers soon began designing for digital media. Today, graphic designers have job titles such as Web designer and UX designer. This is an example of the democratization of technology, where the ability to create typeset copy went from being a profession to work today that nearly anyone can do—yet a new creative industry was also born.

The democratization of manufacturing technology is creating new opportunities that didn't exist previously. The industrial revolution was a transition from making things with our hands to making things with mechanical or electrical machines in factories. A new revolution in manufacturing is happening as digital fabrication technology such as 3-D printers becomes more affordable and accessible. As more and more objects embed electronics to make them smarter, we will have a whole new industry that some call the Internet of Things. We will find new ways to combine art and science, engineering and craft, design and technology.

In the future more people will be creating their own jobs instead of finding a job. The question is not what kind of work you can do but how your work can create the greatest value. We have to be constantly learning new skills and coming up with new ideas, changing as the world changes. How can more and more people have engaging, purposeful, gratifying work, the kind of work that makers do? The maker seems more essential than ever: a sense of agency, self-determination, self-reliance, resourcefulness, collaboration, flexibility, and a can-do attitude.

The creative economy requires more than people who see themselves as creative. It requires people who are productive. They have confidence in their creative abilities, but they have the technical skills and a mastery of process that allows them to develop ideas and turn them into something real, something valuable. Or to solve a problem that others find difficult to address. In short, they are skillful at applying tools and methods to important problems.

The creative economy also elevates the importance of collaboration. There are two kinds of collaboration that we might think of: the product of teams, and the product of communities. Collaboration with a team is tight, intense, directed, and closely managed. Collaboration with a community is loose, open, and distributed. Team collaboration is usually an internal process, while community collaboration is external. If you are a leader of a team, you should be able to get team members to agree and assign them tasks to do. Communities have leaders, but they don't manage people and tasks. You can't really tell others to do something you want them to do, at least not in the same way you might ask a team member. However, you can ask for help and invite others to do something in an open-ended way. The Internet enables community collaboration in an open and self-organizing way, but it's not necessarily easy to do. Learning this new form of collaboration with a community is essential to tap into a new source of creativity and innovation that can expand what individuals as well as organizations can do.

There are many questions that we don't know how to answer: our own future and the future of the planet seems unpredictable. We have so much to learn, and so much of it is learning from each other and combining our individual talents in productive new ways.

10

Making Is Caring

People have asked if the Maker Movement might be turned toward something productive, not just making for the sake of making. This is a common misperception, that somehow makers are hobbyists wasting time on their own projects. Making itself is good and productive. The Maker Movement has grown in leaps and bounds by the very encouragement it offers to people to participate, because making for its own sake is something they'd enjoy. I wouldn't want to critique an artist's work by saying it seems like art for art's sake. There's real value in play; there's value in making. I get what people mean when they want to direct makers to "do good," but what does "good" mean? For me, if making is good on a personal level, it also impacts our friends and family and the larger community. As the maker movement grows and matures, there's much more the maker community can do to broaden its impact.

After visiting Seattle Maker Faire, Ethan Toth, a high school student who saw himself as a creative maker, realized that his own town needed a Maker Faire. So, for his senior project, Ethan produced a Mini Maker Faire for Wenatchee, in central Washington. Ethan got some other students to help him, and they met for two hours after school and after sports to do the planning. He went out into the community to find makers and also talk to the local government and university. A city official he approached had not heard of the Maker Movement until Ethan told her about it. She was intrigued and offered their support.

"Initially, I did it because I love to create things, and I just wanted to share it. From there on, it got taken up by everyone else as well," said Ethan. The whole idea was to generate the kind of excitement in the community that could lead to the creation of makerspaces. "It started out as a simple idea, but with the amount of help I've had from everyone, and the amount of help that my friends have provided, it turned into something incredible." What a great project for high school students not only to get the experience of organizing an event but going out into the community, talking to many people to learn what they do and to get them involved.

In 2015 there were 151 Maker Faires around the world. Many of them were organized by volunteers acting as community organizers. A local Maker Faire helps to build a maker community and create relationships among different makers and groups. Somehow, when you bring all these creative and clever people together, there is a pervading sense of generosity, which comes from each of us sharing our time, energy, and creative work. This generosity is an expression of how we care for each other. It leads to a maker-inspired sense of community service: helping other people help themselves.

TIKKUN OLAM

Hacking, as I've said, is a way to solve problems, often through software. A hackathon is an immersive, collaborative form of hacking, typically taking place over a weekend. A good hackathon promises to accomplish something that has an impact beyond the event. A poor one is no more than a pizza party for geeks or a developer recruitment fair. Hackathons can also be used for hardware development. Sometimes they go by the name makeathon.

In the fall of 2015 I was a judge at a makeathon in San Francisco organized by Tikkun Olam Makers, a group from Tel Aviv, to address assistive technology. Tikkun Olam is a concept from the Jewish faith, literally meaning "repair of the world," and it refers to actions undertaken for the good of society. This makeathon was a seventy-two-hour-long event hosted at TechShop in San Francisco.

Before the event started, the organizers did a lot of preparation to make it successful. First, they defined a set of specific problems that would be presented as challenges to solve. Second, they recruited makers and organized them into teams based on their complementary talents. This meant that when people came to the event, they could just get to work. They didn't have to come up with a new problem to solve, and they didn't have to figure out how to build teams. When you have limited time to make something—and hardware inevitably takes more time than software—coming in prepared is a much better way to make the time productive.

In addition, these weren't problems dreamed up in a whiteboard session. Tikkun Olam Makers had worked with several nonprofits, such as United Cerebral Palsy, to identify one or more "need-knowers": someone who personally represented a specific problem. The challenge wasn't abstract; it was someone's problem that needed solving. Here are some of the projects and the challenges they addressed:

NOW Mobility: How do we take a wheelchair to the next level?

VRAAT: What if virtual reality could help create new experiences for the disabled?

Bridge-It: What if getting in and out of a wheelchair was ten times easier?

Context: What if technology could translate sign language into text?

Doorman Assistant: What if there was an easier way to open doors for a person in a wheelchair?

Smart Ass: What if we could map pressure points for a person seated in a wheelchair?

Grabber: What if there was an affordable grab-and-release device?

Free2Pee: What if female wheelchair users could relieve themselves unassisted?

Carry Crutches: What if a person using crutches could safely carry things around—including an uncovered coffee cup?

The problem was represented by a person who actually experienced it. One woman, named Kim, was born without limbs, and she worked with the Grabber team on a 3-D-printed device that she could use to pick up things at a distance with her mouth. Because she couldn't make it to the first day of the makeathon, the team did a bunch of work without her, only to find that when she showed up, nothing they had created actually worked for her. For example, they weren't aware that there were limits to the weight of objects she could pick up; they had only focused on the mechanics of grasping. The group of makers kept iterating, creating a variety of 3-D-printed grabbers attached to store-bought fishing poles. Toward the end of the event, I asked her if the grabber would really help her. She responded affirmatively: "This will allow me to do more things myself."

I felt like the need-knowers became makers as part of the process. They really adopted the mentality "I can learn to fix things for myself." It wasn't that there are really smart people who can help them; they helped make the process better.

What I liked about the makeathon is that the makers weren't asked to develop a *product,* but instead solve a person's *problem* that no one else was solving. It opened my eyes as to how makers can and do serve others. It helped me recognize that the maker community can use its skills and knowledge to serve other people, and specifically to help people who are not well served as consumers.

The weekend after the Tikkun Olam makeathon, I headed off to the next Maker Faire in Seattle. What struck me about Seattle's Maker Faire was just how smart, geeky, and friendly everyone was. Walking among the makers and their exhibits, I realized how much of what Makers were doing might be considered community service. The makeathon had given me a new lens through which to see the Maker Movement.

There are many people who are making in ways that benefit others in the community, whether it's the Pop-Up Science Museum appearing on a ferry crossing Puget Sound and talking about ocean acidification; a lab for DIY biohacking that gives away DNA samples—extracted from strawberries; the girls from the Big-Brained Superheroes Club who developed a binary counting demonstration with a Raspberry Pi; or the nine-year-old boy who developed a sole insert for shoes to keep people's feet warm in the Arctic. Even the marching band strolling through the event seemed to me to be a celebration on behalf of community.

"Queen" Mary Heacock didn't have an exhibit at Seattle Maker Faire, but she was moving around energetically in her wheelchair, and she came up to me and said hello. "I am extremely disappointed in the poor quality of existing wheelchairs, and they are extremely expensive," she told me. "So I have made a wheelchair that can go from the boardroom to the back country. It's the size of a regular wheelchair but it can clear anything up to nine inches." She said it was what she was going to use to go backpacking.

I asked her how she arrived at the idea that she could create her own wheelchair. She said she got the original inspiration from a friend's fishing net, but she began to think she could do something with her idea after she watched a video of Carl Bass from Autodesk at a Maker Faire in the Bay Area, who talked about Fusion 360 design software. "I was able to learn Fusion over the Web, and I worked with their evangelist and created a complete 3-D model of the chair," she said. Then she found someone locally to make it for her. Afterward, she sent me a video of her on a trail in her wheelchair, which she calls "The Throne."

DIYABILITY

DIYAbility is a term used by John Schimmel, who set up a makerspace in New York City for disabled people to develop their own personalized, customized assistive technology. Joe Olson, whom I met at a Mini Maker Faire in Washington, D.C., in 2014, is a wheelchair-bound maker with DIYAbility. He has been in a wheelchair since 1999 when, just out of a high school, he broke his neck near Lake Superior in a diving accident that left him paralyzed.

After his accident, he spent a period of time in rehabilitation in Colorado and then went to Michigan State University, where he received a bachelor of science in mechanical engineering in 2004, and then moved on to earn a master of science in rehabilitation science and technology from the University of Pittsburgh in 2007. In Pittsburgh he worked as a graduate student at the Human Engineering Research Laboratories, where, he said, "I learned the skills needed to actually build what I only could draw and imagine before." He moved to Baltimore for a job with the government as a facilities engineer in early 2010.

Joe showed me what looked like a small metal mitt that he designed as a replacement for the joystick he used to control his wheelchair. The standard "goalpost" joystick didn't work for him, and he began to develop a prototype with the idea that a handle should match the contours of his hand. It worked better than anything he'd bought, and over the years he has kept refining the design, making it lighter and cheaper to produce, with a comfortable nonslip surface. He decided to launch a company called ErgoJoystick[1] where he designs and manufactures these controllers, believing that others might also benefit from his designs.

"The best-designed things come from users," Joe said. "Would you buy a car designed by someone who doesn't drive?" Joe intends to make new designs for power and manual wheelchairs. Wheelchairs are more than just vehicles used to get around. A person spends most of his or her day in it, doing everything in it, inside and outside the home. In some ways it is an extension of the self, which is exactly "why I need it to do more," Joe concluded.

Nicolas Huchet wrote about building himself a prosthetic hand in *Make:* in January 2015.[2] He had no experience as a maker at all; he just happened to stumble upon an exhibit of 3-D printers in Rennes, France, where he lives. He asked the people running the printers if they could print a robotic hand and was met with classic maker can-do enthusiasm. Someone suggested downloading the pattern for a Robohand and using the 3-D printers to make one.

Robohand got started in 2011 when a woodworker named Richard van As had an accident in his workshop and severed all the fingers of his right hand. He discovered, as Nicolas would, that prosthetics are not affordable for most people: advanced models that have articulated fingers with variable speed and multiple grip patterns and sensors—in other words, that get as close as possible to the functions of an actual hand—cost $50,000 to $80,000. In the United States, insurance may cover a more basic version that costs around $8,000, but in countries like Colombia or Russia, the solution is a hook. In many developing nations, it's nothing at all.

In January 2013, Robohand published the design of its first successful prototype open source, later adding seven additional designs to choose from, as well as video tutorials for making them. People anywhere in the world can download a design and then use 3-D printers and CNC machines to make medical-grade, custom-fitted, lightweight mechanical hands. Wearers can bathe and swim with them. The Robohand pattern has been downloaded more than nine thousand times.

At LabFab in Rennes, Nicolas and his new maker buddies launched the project. "It was truly an international effort," he writes. "Printed digits from Rennes, muscle sensors from the United States, and design input from Brazil. The foundation is a 3-D printed hand, equipped with activators to move the digits and joints, fishing line to connect the actuators to those joints, muscle sensors and a socket where they rest, batteries, and an Arduino brain." Total cost of the Bionic Hand: about $250.

The Bionic Hand received awards for its design in museums and institutes globally. Following this success, Nicolas created My Human Kit, a foundation to support people developing solutions for disabilities and to collaborate with institutions like Johns Hopkins University in Baltimore and the BioRobotics Institute in Pisa, Italy. They intend to extend the range of personalizable and affordable products for people with special needs.

MACGYVER MEDICINE

José Gómez-Márquez, originally from Honduras and with a lab at MIT, recognized that developing countries acquired secondhand medical equipment from the developed world. When it broke, they'd have to fix it without having access to parts. José saw that nurses had to hack the equipment to make it work, and they did so in such clever ways that he dubbed them "MacGyver nurses."

In an article he wrote about DIY medical kits for *Make:* magazine, José talked about the first time he and his team saw a medical hack: "It took us two hours to convince the nurse, Daniela Urbina, to show us how she had fixed the cracked diaphragm of her stethoscope. A young woman from central Nicaragua, she had experimented with various plastics to replace it, and settled on leftover overhead transparency material cut into a circle and taped inside. It wasn't pretty, but it worked."[3]

José started MakerNurse, a program that sees nurses as the lead innovators in the health care system. Sometimes they don't understand that they are innovators and that their clever, improvised solutions are valuable. He also created MakerHealth, a program to develop medical makerspaces inside hospitals. "Making in health occurs when individuals shape, form, assemble, and transform objects with their own hands-on skills and nearby resources," reads the website for MakerHealth.[4]

And what if medical devices could be designed for hacking? José and his team at MIT began developing MEDIKits (Medical Education Design and Invention Kits), construction sets designed to encourage invention among doctors and nurses in the field. José explained:

The kits started as boxes of parts assembled to familiarize MIT students with medical devices, and evolved to include linear components that you can assemble like Lego bricks into a final device. The Drug Delivery Kit was our first experiment. It's divided into core devices: syringes, nebulizers, inhalers, trans-dermal patches, pills, and several other items you might find at your local pharmacy. Then we added modifier elements: color coding, shape coding, couplings, extenders, springs, plungers, compressors, tilt sensors, buzzers, timers, bicycle pumps, and template cutters. We added a healthy amount of consumable general-purpose materials: zip ties, Velcro, adhesives, paper and plastic sheeting, tubing, needles, and respiratory masks.

When users began to snap on, extend, and test their creations, something emerged that we did not anticipate: they hacked our kits. It starts with someone asking permission to simply cut a piece of tubing and bypass our carefully designed coupling. Or taking a part they find in one kit and using it for another, for example, adding diagnostic tubing into a mechanism to disable syringes for safety. As users take ownership of a kit, you as the kit designer become less involved in training people how to use it. So for the kit to be successful, you have to design for hack.[5]

In 2014, at the Maimonides Medical Center in Brooklyn, New York, José and a team from the hospital organized a Mini Maker Faire, the first one in such a setting. Because a hospital runs 24-7, this Maker Faire had a day shift and a night shift. José posted some of the highlights on the *Make:* blog:

> The first thing that rolled into the venue was a six-foot-tall, life-size, animatronic medical simulation mannequin used at the Center for Clinical Simulation. It was part of an exhibit called Remaking the Hospital Bed, where Mini Maker Faire guests could attach ideas of how you could make the typical hospital bed a better one. The simulation environments are critical for health making—you don't want to experiment your ideas on a patient when you can start with a robot.
>
> Victor Ty is an oncology nurse by day and a Lego master builder every other day. He wowed visitors with his recreations of linear accelerators for radiation treatment built with the purpose of helping pediatric patients cope and understand the treatment process in what would usually be a big scary machine in the eyes of some very sick kids. By learning how the machines work and what's about to happen to them, kids, especially those with sensory, language and cognitive challenges, can be more comfortable with treatment.
>
> A staff member from the Child Life Services team demonstrated the various tools they make to help kids understand what the body goes through during different treatments. With a mixture of Karo Syrup, Hot Tamale candies, Jelly Beans, and food coloring in a bottle, you get an amazing representation of the body during an infection and how our blood fights back. I have friends who design advanced simulation systems for Army medics, and this visualization of the blood system is one of the

best (and cheapest) I've ever seen. Oh, yeah, and the kids actually make it themselves. That's health making.[6]

Another project that is hacking health care is the Open Artificial Pancreas System (OpenAPS), an effort to help people living with type 1 diabetes. It's an open-source design for hacking several diabetes monitor and insulin delivery devices together with a Raspberry Pi or an even smaller Intel Edison, which adds up to a rudimentary DIY artificial pancreas. OpenAPS grew out of the Do-It-Yourself Pancreas System (DIYPS), created by two Seattle residents, Dana Lewis and Scott Leibrand, in the fall of 2013.

They are also hacking the health system, moving faster than the federal Food and Drug Administration, using the hashtag #WeAreNotWaiting. Dana Lewis wrote on the blog for DIYPS.org: "We in the diabetes community have seen a series of needs that are not being met with our existing, FDA-approved medical devices that are out on the market. From not-loud-enough alarms to not enabling us to track critical information like temporary basal rate history on the pump itself, these are the needs that drove me (and Scott) to first build DIYPS."[7]

Michelle Hlubinka, whom I've worked with for many years, and her husband, Robert Cook, have a son, Ion, with type 1 diabetes. She explained to me how hard it is to manage diabetes:

> You have to think like a pancreas, because the islet cells don't work as they do in nondiabetic bodies. You manually balance sugar and insulin to stay alive. The "dead in bed" nightmare is something parents of type 1 diabetic kids (like me!) fear: for the ten hours or so your kid is asleep (during which, you hope, you get some sleep too), you have to keep your kid's blood glucose (BG) low enough to avoid long-term effects, like damage to the small capillaries in the eyes, but not so low that you inadvertently starve the brain of sugar, resulting in seizures that, if untreated with fast-acting sugar, could result in a coma or death. It's a crazy, complex system, and my husband, Robert, has had many sleepless nights wrangling Ion's BG.

Robert explained that a biomechanical artificial pancreas, as opposed to a biological one, has three components: an insulin pump, a continuous glucose sensor, and algorithms. "The goal is to take the periodic readings from a glucose sensor and adjust insulin dosing to control blood sugar to a healthy range. This is a hard problem because dosing insulin is so dangerous."

The device is clumsy and a bit awkward, with dongles and cables connecting the parts and a five-volt battery pack. "Since this is not super portable, I imagined using it just during the night for Ion, but the adults using this carry it around with them inside bags or backpacks and seem happy to do so since it relieves them of a major burden. That is the burden of hour-by-hour attention, of sudden crises, of shortened lifespans. It's why people are working so hard on this." A working artificial pancreas, Robert said, would be the most important advance for type 1 diabetes since the discovery of animal insulin almost one hundred years ago.

HUMANITARIAN CHALLENGES

Dara Dotz is the cofounder of Field Ready, a nonprofit that provides disaster relief through on-demand manufacturing where supply chains have been broken. Field Ready has responded to earthquakes in Haiti and Nepal, deploying 3-D printers and small-scale injection-molding systems to help make repairs in places where spare parts are tough to come by.

Dara has been involved in aid work since the age of fifteen. She became an industrial designer, but she liked "the immediate feedback loop" of solving specific problems for people. She visited Haiti two years after its 2010 earthquake, without any specific idea of what to do, but once she was there she noticed that there were radical supply-chain problems. "People couldn't get access to what they needed," she said. In looking at water systems, she saw that parts were missing or replacements couldn't be found, and the local people weren't trained in maintaining or even fixing wells. Dara started wondering if 3-D printing might be able to solve the problem. Then she made friends with a local nurse.

Dara explained:

> She had delivered five babies in one night, and she ran out of medical supplies. She cut the fingers off her own latex gloves to tie off the umbilical cords of infants to prevent neonatal sepsis. That meant she was delivering babies with bare hands, exposing her possibly to HIV and risking her own health. Upon hearing this, I thought: *I bet we could 3-D print umbilical cord clamps.* I put out a call for an old 3-D printer.

Although Dara was a designer and had been in TechShop in San Francisco, she hadn't used a 3-D printer herself. Soon she learned how to use a 3-D printer in a container in Haiti. "We came up with a prototype, and then we had some nurses test it before we finalized the design," she said. It evolved over the course of two years. They trained local community members to use CAD tools and 3-D printers so they could produce the parts locally. Eventually, the clamps were placed in birthing kits, to be sent into the mountains where women give birth without doctors. "Field Ready wants to focus on more localized, small-batch deployments," said Dara. "We want to solve problems in the field with small teams."

I saw Dara at World Maker Faire with Lisa Marie Wiley, a woman with a fierce look in her eyes. She wore jeans cut off at the knee, and her lower left limb was a black 3-D printed prosthetic that she had designed with Dara.

"I might not look like a soldier to you, but I am," said Lisa Marie. She explained that she had fought in Afghanistan and lost her limb after stepping on an IED. Because she was small in stature, existing prosthetics did not fit her well, and they were expensive. With Dara, she designed a better prosthetic that could be 3-D printed for under $20. Her wish was that she could avoid talking about her prosthetic during the first five minutes of talking with someone. Her new custom prosthetic helps her start a different kind of conversation.

ENVIRONMENTAL CHALLENGES

Tim Dye is an air-quality scientist. He has traveled the world measuring air quality in cities like Shanghai and Mexico City. The company he works for produces a website called AirNow.gov that reports on air quality in cities. He and Michael Heimbinder, an educator from New York City and the founder of HabitatMap.org, are collaborating on a project called AirCasting.

AirCasting follows the model of another open-source project, Safecast, which was a response to the Fukushima earthquake of April 2011. The goal was to measure and map the radiation level in the affected areas in Japan. One problem was getting enough Geiger counters, which were much in demand. The other was creating an independent source for data, because some believed the government was underreporting the radiation levels so as not to cause alarm. Bunnie Huang created a reference design for a compact Geiger counter named bGeige. It was small enough to be carried around by a person and take readings every five seconds. Another group at Tokyo Makerspace created a Geiger counter that could be mounted on a car, enabling far more readings in many locations. Safecast was an open-source hardware device that could transmit data to a smartphone, so that the person using the device could visualize the sensor data. It would also be transmitted over the Internet to a site that aggregated the data and generated a detailed map of the recordings.

AirCasting followed this model to crowd-source air-quality data from far more sources than are practical today. Until Tim explained it to me, I didn't realize that our monitoring of air pollution was so limited. For instance, in Santa Rosa, California, there is one device measuring air quality for the whole area. It is managed by the federal Environmental Protection Agency, which probably pays a lot for each monitoring station and can't afford to have many of them. In addition, the location of a monitoring station can be determined by politics—to avoid areas that might have high readings. We know air quality varies based on proximity to roads and other sources of pollution. Having more places reporting is the goal of AirCasting, which makes the data it collects available as well.

AirCasting could help people evaluate their own exposure to pollution and take action.

Tim is also a maker, and on his own time he compares DIY devices for monitoring air quality to professional-grade monitoring equipment. He loves the idea of a cheap monitoring device, but he wants to know how accurate each of them is. Like cameras, the higher the quality, the better the resolution of the sensor data. However, that doesn't necessarily mean that the higher-priced device is better. Tim tests these devices in his garage in a rather unique way: with the garage door down, he lights a single piece of paper and lets it burn. Then he checks the readings of the homebrew devices against well-calibrated professional equipment.

Tim is interested in getting schools involved, especially in developing countries. AirCasting is a way to connect kids to what is happening to the air they breathe. Having a monitoring device is like having a weather station at a school, which is a good thing as well. But the air-quality data can lead to direct action to improve the air quality in a particular area. Students could become advocates for a community's right to clean air, and carry with them a data set as evidence of the problem. The data can also be compared with public health data to see if there are correlations between bad air and poor health.

"Air quality is really scary. Oftentimes, the AQI [air quality index] can be abstract," said Bernie War, a maker originally from Miami who has been living for several years in Shanghai, a city with serious air pollution. He told me he liked everything about Shanghai, except the air. He first developed an open-source air purifier and then created an air quality monitor called the Ella Assistant that helps a person act on data about air quality or weather. "We wanted to develop something that provides you with a simple suggestion, such as you should wear a mask, you should close your windows, or if you should run or not." Bernie designed the Ella Assistant to function independent of a smartphone. You could put it on the countertop or mount it on a wall so it can remind you of the right action to take.

The OpenEnergyMonitor was developed by Trystan Lea from Wales to help us understand the amount of energy we use. He explained:

We can't see energy. We might see the wind turbines, but we don't know when our power is coming from them. How can we make the data something people can relate to? In early 2009, in my last year of university, I bought an off-the-shelf energy monitor, and it wasn't Web connected. I couldn't see any historical data. I'd been using Arduino for about three years, and I thought I'd try to build an open-hardware, open-source energy monitor that was Web connected.

He believes in the benefits of open-source technology: "At the moment it's hard to control your heat pump depending on what your solar panels are doing because the protocols for it are closed, or you can't access the firmware, so you can't change it," he said. "Open source gives you ownership over technology and the empowerment to develop and produce technology for your own life, which is so positive."

My collaborating author, Ariane Conrad, met Trystan during an event called Proof of Concept 21 (POC21), a five-week innovation camp for prototypes of open-source solutions in the areas of sustainable energy, food, and mobility. The camp, which took place on the grounds of a derelict castle outside Paris, united people who usually work separately, like designers, engineers, scientists, and storytellers, with innovators from the open-source and Maker Movement. The name POC is a play on COP, the United Nations' annual meeting on climate change, which in 2015 took place in Paris for the twenty-first time. After twenty years of these meetings, carbon dioxide levels have doubled, so this group of concerned makers and designers took matters into their own hands.

Simon Kiepe, one of the camp's organizers, said: "With POC, the proof of concept, we are actually doing what needs to be done with our own hands. It's basically the contrast between talking and doing. I think people feel frustrated by twenty years of talks with no real results, and they really want to start doing something."

One project at POC21 was the Showerloop, developed by Jason Selvarajan from Finland. Showerloop is a hot-water recycling system that uses ten times less water and heat energy compared to a normal shower. Jason said that "it essentially means you can shower for as long as you want." Showerloop captures, pumps, and filters water through several stages. "When the water comes back through the shower head, it's clean and bacteria-free, and still warm," Jason explained.

According to Jason, a normal shower uses 10 liters (2.6 gallons) of water per minute, with the average shower lasting ten minutes or so. That means it's around 100 liters of water per normal shower. "Showerloop is more like 10 liters of water," said Jason. "It's ten times more efficient in terms of water and energy use." How does it work? "There's a pump connected to the drain. Instead of the hot water running down the drain into the sewers, it is pumped up through a series of filters. There's sand to remove particles. Then activated carbon removes chemicals and smells. After that, we use ultraviolet sterilization to sterilize the water so it's drinking quality." Jason told Ariane:

> I'd really like to do work with refugee camps and emergency scenarios. In situations where utilities aren't working anymore, or where the grid's down, you need to have a low-cost sanitation or water treatment system. Hygiene is a huge issue: If people don't have proper sanitation, they get sick. The next thing I want to do is include the whole water system in a home: connect sinks, toilets, dishwashers, and washing machines. We save even more energy and water, and then we're more resilient, much more adaptable to anything that would come at us.

A huge fan of science fiction, Jason said that he always wanted to be an entrepreneur and have a positive impact. "It's not so easy because I'm not rich, but I've also kind of discovered that you find what you need."

Also at the camp was Dawn Danby, a designer who works at the Autodesk Foundation, which supports the use of design for addressing social, environmental, health, and education challenges. The foundation calls this "impact design." Dawn said:

> We are building up the capacity to respond to disruption, to respond to changing economic and ecological circumstances. We need the maker spirit applied to everything around us

because the big systems that support us are not doing as well as they have in the past. My advice to anyone who wants to do sustainable design is: don't be afraid to get out into the community and start asking questions. Go and figure out who in the world has already done something like what you are doing. There are tons of solutions out there. We can share and learn from what other people are doing. Get in touch with those people. Figure out what they did in that country that you can bring to yours.

Reflecting on the larger meaning and potential of what happened in five weeks at POC21, Ariane wrote on Medium:

> For so long we've been so dependent. We don't know how to meet our most basic needs without the global, exploitative system. Most of us don't have the means to produce our own energy or our own food locally. We've become frightened of this dependency, yet helpless to break out of it, powerless to change such massive systems.
>
> DIY solutions give us a taste of empowerment and agency. When we start getting a handle on producing our own food and our own energy (and next, our own housing, our own tools, our own vehicles and communications devices, and then our own governance and finance systems), we stop feeling so dependent, and we become less afraid.
>
> That is the true meaning and potential of POC21, even if it doesn't solve all our super-complex global problems in these five weeks. It's about people participating in the solutions.[8]

REACHING THE UNDERSERVED

A contemporary of John Dewey, Jane Addams focused on the human needs of life in crowded, industrialized Chicago. She became a social activist and advocate for the poor, devoted to improving the daily lives of immigrants who worked long hours in factories for low wages and lived in crowded conditions. Life was unsafe, unsanitary and unhealthy. Addams went to live with the immigrants in their neighborhood, establishing what she called a Settlement house known as Hull-House. Addams was hands-on and she worked inside the community to observe the patterns of life and look for ways to improve the lives of those who lived in poverty.

She called Hull-House "an experimental effort to aid in the solution of the social and industrial problems which are engendered by the modern conditions of life in a great city."[9] Addams created associations, clubs, and social centers, turning "disused buildings into recreation rooms, vacant lots into gardens,"[10] and she established medical clinics and schools. She was an organizer of spaces and a developer of innovative services. She began to invent the social fabric of a city that lacked a safety net for its poor.

She saw what was happening to children of immigrants whose parents were away at work, and she created activities and a place where they could come every day instead of being locked inside or roaming the streets. She believed in the power of recreation as much as education for adults as well as children, so she organized hikes and built swimming pools as well as establishing debating clubs and poetry readings.

In her book *Twenty Years at Hull-House,* Addams describes her many social experiments such as methods for trash storage and removal, improving the diet of immigrant families whose working mothers had not the time to cook nutritious food, and protecting children and young women from exploitation. Addams consistently described her work as experimentation: "We continually conduct small but careful investigations at Hull-House, which may guide us in our immediate doings…. Some of the investigations are purely negative in result."[11] Addams did not start out with a fixed set of solutions. She created a place where she could invite the community in, and it sounds a lot like how we might think of makerspaces today.

Addams expressed her vision of the community she sought to create: "In a thousand voices singing the Hallelujah Chorus in Handel's *Messiah,* it is possible to distinguish the leading voices, but the differences of training and cultivation between them, and the voices in the chorus, are lost in the unity of purpose and in the fact that they are all human voices lifted by a high motive."[12] Active participation builds community and democracy.

Colin "Topper" Carew, who was an architect, and a writer and producer in Hollywood before landing at the MIT Media Lab, has been working with historically black colleges and universities (HBCUs) to help them provide opportunities for students to make and invent. Carew said:

> The Maker Movement is one of the few things out there that has the possibility of leveling the playing field in technology. There is this that African Americans don't belong in technology, but we've got to reverse that thinking. By creating opportunities where people can make and do, we can get people into the tent, and once they get in, invariably they get excited. Getting hands-on experience seems to be the big difference. One of the great detriments to the inclusion of more African Americans and Latino Americans in technology has been a lack of access. Makerspaces provide that possibility. In Boston, there are a significant number of makerspaces and the African-American population does find its way to them.
>
> I see making as a movement, and we recognize that change is inevitable. If we understand that the African American and Latino communities have a great deal that they can contribute long-term to American productivity and prosperity, to economic development, to quality of life, people can get excited. It's important, but I don't think there are enough people that understand that yet.
>
> All movements start small. We are in the early days, the beginning of a movement that will move African Americans and Latinos into the innovation and invention space. Making should be an opportunity we provide to all people. Starting at very young ages, we have to grow and nurture our young people.

Schools will eventually come on board, and we need them, but most of the modeling happening in this country right now is going on outside of the schools. This is where young people get turned on.

By creating an environment where young people can play, experiment, and make mistakes, where they can learn to solve problems, where they can flourish artistically, where they can use code just like a paint brush or a musical scale—all of these things are what make for a more complete and well-rounded individual. Ultimately, they become more confident, curious, and capable.

One of the talking points I use with HBCUs is to look at what happens when the African American community gains access to tools. Go back to the early days of miniaturized recording machines. A young person today can have access to a recording device, which at one point in time might have cost $400 an hour. Now they had a machine. I watched as these young people got the means of production, and with the code being the twelve notes of music, they began to invent things. One of these things became rap, one became hip-hop. Those things are now Americana. That culture has impacted everything.

Carew's comments reminded me that at World Maker Faire, I had met one of the old-school DJs, hip-hop producer Jazzy Jay, one of the founders of Def Jam Recordings. In the early days of hip-hop, Jazzy Jay was building a home recording studio because nobody could afford a professional sound studio. He said in an oral history interview by NAMM that he started a production studio with "the crappiest stuff you could dig out of the garbage, out of people's stereos that they threw out." He added that "it was all primitive, but it taught me how to manipulate electronics. We couldn't afford going to the store to pay for this stuff."[13] At Maker Faire, Jazzy Jay was enjoying himself, appearing with DJs from Thud Rumble. "I get this event," he said to me. "I'm a geek myself. I love seeing all of this." Carew said:

Makerspaces can give young people the same kind of access to tools for production. Once they have access and the code is understood by those communities, there's going to be an explosion—I don't know what it is. But I know that something wonderful is going to happen and it will be characterized one day as Americana. So the value of propagating a movement that brings those populations into a movement is going to have a wonderful, long-term, highly productive impact on American culture, productivity, and competitiveness.

This is beginning to happen in youth makerspaces across the country. The Maker Education Initiative looked at fifty-one youth makerspaces in the United States and found that average participants were forty-two percent white, twenty percent African American, eighteen percent Latino, and fourteen percent Asian. Individual makerspaces may vary, for instance, as some may be predominantly white and some may be predominantly black.[14] Carew continued:

When I talk to HBCUs, I help them see the vision and see what is happening. At Spelman College, young women designed a camera that mounts on a laboratory microscope. They designed and fabricated it, and the new camera replaced an $850 camera. The expensive cameras now sit in a pile on the floor. Now every student who wants her own personal camera can get one for $20. That's why the Maker Movement is an important platform for long-term productivity and accessibility.

The Maker Movement has the potential to transform how the poor see themselves and how we see them. If nobody expects you to be a maker, how would you ever discover these talents? John Bare from the Arthur Blank Foundation in Atlanta once made the point that in Atlanta, as a male child, you cannot go without having many opportunities to play football. He asked the question, how could that be true of making? In other words, our society thinks that the poor, particularly African Americans, should play sports and have every opportunity to do so as a way to keep young people in school and become productive citizens. The community, in turn, takes pride in its sports programs and its athletes.

What if we looked at the same group of children and believed as strongly that they could be active, creative learners who stayed in school because it was a positive experience for them and because it helped them become the person that they wanted to be? What if they were able to develop valuable technical and creative abilities in a collaborative environment?

We must realize how much talent is being wasted by the limited resources we devote to support the growth and development of young children in poor communities. By not discovering and developing that talent, we not only miss something of great value, we also prolong the pain and suffering not just of individuals but of entire families. It's like having a unique natural resource buried in a big mountain: If we understood how valuable the resource was, we'd spend a lot of money to extract it. Becoming a maker is not just a set of recipes or a set of tools, not even a list of DIY virtues. Making is a practice that promotes a person's social and emotional well-being while developing skills that make a person useful and productive. A sustained practice is a set of ongoing experiences that build on each other in the context of a supportive community. It requires people who care deeply.

Irwin Kula is an eighth-generation rabbi who is head of the National Jewish Center for Learning and Leadership. In a talk he gave in 2014 at the Business Innovation Factory conference, he said: "Religion is not about creed, dogma, or tribe. First and foremost, religion is a toolbox designed to help human beings flourish. Religion's just a technology. How the hardware of humanity gets used will depend on the software."[15]

Making is not a religion, per se, but we can use making to help "human beings flourish." Making is a way, both simple and complex, through individual practice and group support, that we may be able to gain wisdom to appreciate the unique gift of our own life, show compassion, and care for other people.

Rabbi Kula and I are on the board of Big Picture Learning, a network of urban schools developed by two educational renegades, Elliot Washor and Dennis Littky. Big Picture Learning schools offer an individualized approach to learning that combines work in school with internships in the community. This approach is based on listening to the students and then helping them find their way.

The Communications Director of Big Picture Learning told me that as a child he grew up in foster homes and bounced from one school to the next. What made a difference for him was that when he went to Big Picture Learning, a counselor asked him a question that he said nobody in his life had ever asked him before: "What are you interested in?" He replied that he liked to draw, and nobody knew that about him. His counselor suggested taking a class in graphic design, and the subject clicked with him. Big Picture's secret formula is caring for the students and helping them express themselves, which allows them to flourish as human beings.

The Maker Movement asks us not only what technology can do but what good people can do as a community to use such tools to take care of ourselves and each other. We can fully engage our humanity in shaping science and technology, not just allowing it to follow the cold logic of systems, bureaucracies, or markets. We can put humanity at the center of our future once we understand that caring is in our toolbox.

11

Making the Future

I started this book by saying that "nobody needs to make anything today." If that were really true, we would not have much of a future. We'd be like the well-fed, well-cared-for characters on the spaceship *Axiom* in the Pixar movie *WALL-E*. In one scene, the captain is arguing with Auto, the computer, about returning to Earth:

> **CAPTAIN:** *Out there is our home. Home, Auto. And it's in trouble. I can't just sit here and- and- do nothing. [moves back toward Auto] That's all I've ever done! That's all anyone on this blasted ship has ever done. Nothing!*
> **AUTO:** *On the Axiom, you will survive.*
> **CAPTAIN:** *I don't want to survive. I want to live.*
> **AUTO:** *Must follow my directive.*

Ah, the ultimate consumer lifestyle, sitting around and being waited on by robots and following their programmed directives. Is that our future? The scene reminds me of when I was in the hospital as a child. I had people taking care of me; I had a black-and-white TV and a room to myself. However, I longed to be active and do things that other kids were doing, like riding bicycles. I wanted to be able-bodied so I could get out to go play and live.

Like the captain said, humans "can't just sit here and do nothing." We have to make and do things to live. We have to be active and engaged to be fully alive. We need a deep sense of purpose that compels us to act. We need to participate as well as cooperate with others to enjoy life and be productive. There's a future to make.

THE UNISPHERE

If there's an inspiring icon for the Maker Movement, it's the Unisphere. A twelve-story-high stainless steel sculpture of our planet in Corona Park in Queens, New York, it was the symbol of 1964 World's Fair. The Unisphere was "commissioned to celebrate the beginning of the space age" and dedicated to "man's achievements on a shrinking globe in an expanding universe." The Unisphere represented the theme of global interdependence.

The 1964 World's Fair was intended to show us what the future would hold for us, as presented by companies like Disney, Westinghouse, IBM, and General Electric. General Motors' attraction was the Futurama 2 ride, a tour of the future with a boldly confident narrator: "Welcome to a journey into the future, a journey for everyone today into the everywhere of tomorrow. Let us explore together the future, a future not of dreams but of reality." Beginning with humans colonizing the moon and ending with the cities of tomorrow, the narration concludes: "Technology can point the way to a future of limitless promise, but man must chart his own course into tomorrow, a course that frees the mind and the spirit as it improves the well-being of mankind."

I came across an essay written by science fiction writer Isaac Asimov fifty years ago after he visited the 1964 World's Fair. Asimov wrote:

> What is to come, through the Fair's eyes at least, is wonderful. The direction in which man is traveling is viewed with buoyant hope, nowhere more so than at the General Electric pavilion. There the audience whirls through four scenes, each populated by cheerful, lifelike dummies that move and talk with a facility that, inside of a minute and a half, convinces you they are alive.[1]

Asimov asked the question, "What will life be like fifty years from now?" and hypothesized that:

- Electroluminescent panels will be in common use.
- Gadgetry will continue to relieve mankind of tedious jobs.
- Robots will neither be common nor very good, but they will be in existence. The IBM exhibit at the present fair has no robots but it is dedicated to computers, which are shown in all their amazing complexity.... If machines are that smart today, what may not be in the works fifty years hence? It will be such computers, much miniaturized, that will serve as the "brains" of robots.
- General Electric will be showing 3-D movies of its *Robot of the Future*, neat and streamlined, its cleaning appliances built in and performing all tasks briskly.
- The world of fifty years hence will have shrunk further.
- Much effort will be put into the designing of vehicles with "Robot-brains"—vehicles that can be set for particular destinations and that will then proceed there without interference by the slow reflexes of a human driver. I suspect one of the major attractions will be rides on small robot-icized cars which will maneuver in crowds at the two-foot level, neatly and automatically avoiding each other.
- Only unmanned ships will have landed on Mars, though a manned expedition will be in the works, and the Futurama will show a model of an elaborate Martian colony.
- The fair will feature an Algae Bar at which "mock-turkey" and "pseudosteak" will be served.

- The world will have few routine jobs that cannot be done better by some machine than by any human being. Mankind will therefore have become largely a race of machine tenders.
- All the high-school students will be taught the fundamentals of computer technology, will become proficient in binary arithmetic, and will be trained to perfection in the use of the computer languages.[2]

Asimov was prescient, seeing the "wonderful" future even beyond fifty years, including autonomous cars and probes landing on Mars as well as the limited abilities of robots. Yet we might also be frustrated that the world has not kept pace with all his predictions. We are still working too hard at "routine jobs"—why does anyone still have to do chores? And sadly, too few high school students are learning programming languages, although in January 2015, President Obama announced a computer-science-for-all initiative "to empower all American students from kindergarten through high school to learn computer science and be equipped with the computational thinking skills they need to be creators in the digital economy, not just consumers, and to be active citizens in our technology-driven world."[3]

Asimov concludes his essay with a paragraph that I find stunning:

Mankind will suffer badly from the disease of boredom, a disease spreading more widely each year and growing in intensity. This will have serious mental, emotional, and sociological consequences.... The lucky few who can be involved in creative work of any sort will be the true elite of mankind, for they alone will do more than serve a machine.[4]

Are the "lucky few" makers? Is it the fate of the vast majority of people to suffer unrelenting boredom, and isn't our boredom a by-product of consumer culture? Having all the conveniences relieves us of having to do anything too hard. Being told what to buy and how to feel about it distances us from involvement in the process by which things are made. We live in a culture that programs us, and we learn to comply. We don't know how to take control of lives, and we often lose interest in doing anything.

At age fourteen, Joey Hudy, from Arizona, visited the White House, representing Maker Faire in a celebration of student Science Fair winners, and gave the President his business card, which read: "Don't be bored. Make something." The disease Asimov is pointing out is not just boredom but depression, as discussed in chapter 7. Joey is prescribing a treatment.

We have held World Maker Faire for six years at the New York Hall of Science on the grounds of the 1964 World's Fair. I think of Maker Faire as a descendant of the World's Fair, but it's more a People's World's Fair. I think it does give us a glimpse of the future; it could be said to celebrate the promise of technology but features what people and small community groups are doing.

As I wrap up this book, we have produced fifty editions of *Make:* and ten years of Maker Faire. There were 151 Maker Faires in 2015. I wouldn't have predicted with issue number 1 that the magazine would inspire the Maker Movement or that Maker Faire would serve as its catalyst. I just didn't think that far ahead. There was too much work to be done to get out the first issue and prepare for the next one. To be truthful, it was easier to imagine it failing.

It's hard to imagine the future and the big leaps that come, often unexpectedly. In a conversation with Nick Pinkston of Plethora, I asked him why it's so hard. He thought that most people find it easier to imagine the world ending rather than it being something different than what it is. He may be right. For most of our history, we had faith that the future would be better. In the 1940s public opinion researchers started asking American parents if they thought their children would be better off in the future. For fifty years, the majority of those asked said yes. Sometime in the late 1990s, however, public opinion shifted from optimism to pessimism. By 2014, seventy-six percent of Americans polled did not feel confident that life for their children's generation would be better.[5] The media called it "the end of American optimism."[6] It's not just Americans, either. In 2014, for their Global Trends report, the Pew

Research Center found that the majority of people they asked in eight out of ten of the world's wealthiest economies believe that younger generations will be worse off, with the French, Japanese, and British particularly pessimistic.[7]

What I take away from makers I've met all over the world is that if we want life to be better in the future, we need to make it ourselves. It won't be just our making things that creates the future; we have to create a new culture that reestablishes making as something that all of us can do. Maker culture represents a value shift from competition to collaboration, from proprietary to open, from institutional power to individual empowerment, from apathy to agency. It is a learning culture that values practice over theory, experiential learning over textbooks, and creativity over standardization. It will be a network culture, widely distributed and self-organizing, working cooperatively on both global and local levels. We need to be prepared as makers for a future that will require us to be creative, interdisciplinary, resilient, and agile.

The Maker Movement is still emerging, and we are early in its development. Change happens both suddenly and slowly, like making something: the idea of what to make can come to you instantly, well-formed and clear, and yet to make it a reality requires determination, hard work, and time.

SOCIAL MOVEMENTS

I've wondered what other kind of movements might be similar to the Maker Movement. A movement can be defined as a series of organized activities by people working concertedly toward some goal. We often think of movements for social change, such as movements for civil rights, women's rights, and gay rights. Those movements sought legislative, institutional, and cultural change as shared goals. However, I have to pause before considering the Maker Movement in light of these land-mark social justice movements. They are the gold standard. What I see in them, however, is that they develop over a long time, arising from grassroots efforts of people truly committed to change, even a kind of change that they might not experience for themselves, yet a change they believe is inevitable. As Martin Luther King Jr. said: "The arc of the moral universe is long, but it bends toward justice."

Is there a moral imperative driving the Maker Movement? What is driving other movements? I see similarities in the Slow Food movement, which developed in Italy and California as a response to fast food. Slow Food advocates wholesome, local food over processed food with dubious ingredients and obscure origins. The Slow Food movement seeks to develop alternatives to the industrial system of food production and distribution, which is optimized for speed and efficiency. It encourages us to slow down, enjoy the simple pleasures of life, and make connections to real people creating real food. When we become more involved in the process of bringing food to our table, then we have a positive impact on the local environment as well as the local economy.

I see makers also exploring similar alternatives to the consumer culture. DIY is essentially the slow way. To do it your own way allows you to optimize for values that are important to you. You can choose to put uniqueness ahead of efficiency. It is the sum of very personal choices that makes the work of an artist or craftsperson unique.

The work of makers might be considered "slow made." A slow-made object is one whose maker has inserted himself or herself into the process. *Slow made* could be used the way the term *handmade* is used but without insisting that we ignore using machines to make things. Slow made could celebrate the human effort—a combination of manual and mental labor—that goes into creating something. Whether it's building things from scratch or from a kit, or taking an idea all the way from design through build, you shift from consumer to producer, creating value as you produce things for you and your family.

In looking at different kinds of movements, I found myself studying marathons. There are more people running marathons today than ever before, and it has not gotten any easier over the years. More and more people are choosing to do something that is not easy to do, but by doing it they reach an important personal goal. As someone who has never run a marathon and has not one ounce of confidence that I could ever do so, I can't get over how hard it is to run a marathon.

The Boston Marathon is the nation's first and most prestigious marathon. The Boston Marathon started in 1896, inspired by the Olympic Marathon. For most of its first hundred years, only a few hundred people participated. They were very dedicated amateurs, but the field was small. However, a rapid increase in the number of people entering the marathon started in the late 1960s and peaked at the hundredth anniversary of the race. So many people wanted to participate that race organizers had to limit the number of entrants.

NUMBER OF PARTICIPANTS IN THE BOSTON MARATHON, 1960-2011

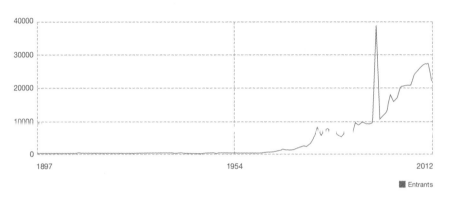

Runner's World started as a magazine in 1966, and a "running boom" occurred in the 1970s and 1980s that engaged twenty-five million people. The first running shoe was introduced in 1938. Much later, Blue Ribbon Sports got its start in 1964, and Bill Bowerman's waffle trainer (a maker story of its own) was introduced in 1974. That company took the name Nike in 1978. The company's stated its mission in advertising that read:

> To Bring Inspiration and Innovation to Every Athlete* in the World.
> *If you have a body, you are an athlete.

Marathons were started by and for an elite group of amateurs, some of whom later became professionals, but the rise in participation in marathons was driven by amateurs. The real change was that more and more amateurs started running for personal reasons. The amateur raced not hoping to win the race, but often just to finish the race—and set a personal goal.

Here are some of the factors that I think changed participation in marathons: First, the number of races in cities increased. You didn't have to go to Boston because the Boston Marathon race was replicated in many different places. Today there are about 1,200 full marathons around the world. Second, marathons became more open and inclusive. Women were allowed to participate in 1972, and by 2009 there were ten thousand women entered in the race. Third, better training that was easily accessible made it possible for more people to participate. Track clubs began offering training programs that made this offer: "If you are willing to train, we can train you to run a marathon." You didn't have to prove that you were a "born" runner. You just had to be willing to train. Also, you didn't have to train alone. You could find a group of people who wanted to train like you, and this group would have a trainer or coach. Better coaching and preparation have helped more people meet the challenge.

The parallels to the Maker Movement are strong, especially as seen through the lens of Maker Faire. Maker Faire is a marathon of sorts for makers, the large majority of whom are amateurs. Events offer a structured form of participation as well as recognition for the participants. Maker Faire shows clearly that many people are interested in what is happening, and not just the makers but what marathons would consider spectators.

Women have often been excluded from making, or at least not encouraged the way boys were, and we know that such encouragement makes a significant difference. More needs to be done to support the practice of making among girls and women. It's also important to make sure our idea of making is inclusive of all forms of making in our community and of all the places where making occurs. Increasing the diversity of participants is crucial for the continued growth of the Maker Movement.

The Maker Movement needs better and more accessible training for aspiring makers everywhere. Makerspaces can provide access to training and create a supportive environment where more people can learn to do something that they might have thought was too hard for them to do. They can also help makers get better, improving their ability to create and make something, often by joining forces with others who share a common interest. Makerspaces provide access to resources, but they also provide a meaningful context for makers to grow and develop. We should seize the momentum that has been gaining around makerspaces to ensure that they are available in low-income communities as well.

No doubt the Maker Movement is connected to other changes that are taking place in our society, just as the growth in marathons reflected not only the personal goals of participants but also a broader societal change in health and fitness.

What we eat, how we exercise and why, what we buy, and what we choose to make are all changing our culture in important ways. A movement is just another way of saying that each of us can be a part of positive change. Many successful movements rise up from independent but connected efforts; they are spurred not by abstract goals but by every-day outcomes such as a better meal, a healthier body, a more considered purchase, or the reward of doing something ourselves because we can. It changes our own lives and ends up changing our culture.

PARTICIPATORY CULTURE

Burning Man, Maker Faire, hackathons, and even Punkin Chunkin are examples of participatory culture. According to media theorist Henry Jenkins, participatory cultures have relatively low barriers to artistic expression and civic engagement, strong support for creating and sharing one's creations, and some type of information mentorship whereby what is known by the most experienced is passed along to novices. Members believe that their contributions matter, and feel some degree of social connectedness with one another (at least they care what other people think about what they have created).[8]

I once went to Detroit Soup, a communal dinner in the downtown area that was organized by volunteers. Each person put $5 in a pot at the door, but not to pay for the food. At the end of dinner, three local people pitched a project that they wanted to do and then people voted—they actually had voting booths. The winner took home all the donations, using it to start an art project or perform a community service. The money was seldom enough to cover the full cost of doing the work, but it was a sign that members of the community were behind the project and supported the artist. I liked the way Detroit Soup combined the act of sharing a meal with funding creative work in the community. It was like Kickstarter but simpler, personal, and more participatory because it was local.

One of the images I use in my talks is a photo of a 3-D-printed Japanese character that I took at Maker Faire Tokyo. Japanese makers impress me by how serious they are about play. After a recent talk at Microsoft, one of the executives asked me if I knew who that character was. I said I didn't. "That's Hatsuna Miku," he told me. "She's a vocaloid." Her name in Japanese means something like "sound of the future." Hatsuna Miku is an animated vocalist, produced by sampling a voice to create a specific sound bank and then using it to generate a song from a Yamaha synthesizer. What makes her an example of participatory culture is that she performs songs that her fans write for her. Fans produce their own videos of her singing their own songs, and they have become hugely popular. Although she performs live concerts appearing as a hologram, and her Facebook page for a 2016 North American tour has over 2.3 million likes, Hatsuna Miku doesn't exist—except that her fans bring her to life through their own creative acts.

Another similar phenomenon is the rise of impersonators and tribute bands. The U.S. Census counts eighty-four thousand people who say their occupation is Elvis impersonator. Tribute bands imitate the dress and sound of groups like KISS. It's the fans taking over for the band, amateurs playing pros. In an interview with *Ear Candy*, members of a Led Zeppelin tribute band named ZOSO were asked:

> *Do you think that the rise in the number of tribute bands, be it The Beatles, Stones, KISS, Zep, or The Doors, is a sign that something is "missing" from today's music?*
> **ZOSO**: *YES! It's too corporate. Creativity is gone.*[9]

I love the response, even not knowing if he means to be ironic. How is it any different than a community theater production of a Shakespeare play? Or cosplay? Tribute bands have defined a context in which to express themselves and connect to other people. What is essential to music is making a real connection between the performer and the audience.

You might wonder what the connection between Hatsuna Miku and tribute bands is, or between Maker Faire and Burning Man, Tesla coils and knitting. For some, there isn't one. But seen differently, there are simply the connections that you make yourself. People who distill Burning Man into a few words will never help anyone understand Burning Man, let alone navigate it. Similarly, Maker Faire is not about a unified message or theme. When people ask me for a tour, I have to explain that I don't have a set path myself. I lead them randomly through ten or so exhibits. I point out that they have seen very different things together, and they all seem to belong. I advise them just to go on their own and create their own experience. Both Maker Faire and Burning Man are unique sets of different experiences that you may or may not integrate into a whole.

Another way to look at it is to consider the contrast between Disneyland and Maker Faire. They are different states of mind, rather like the difference between leisure and play, between comfort and passion. Disneyland packages many different things to make them all look pretty and appealing—in other words, pretty much the same. Disneyland is meant to be predictable, telling you a story that you already know. It's meant to be as familiar as a bedtime story, like Pinocchio, bringing to life the fairy tale of a puppet brought to life. You are meant to like it and not think too much about it. It's a great example of convergent thinking, where the path leads us all to the same place.

Maker Faire, like Burning Man, is the opposite of Disneyland. Parents tell me that when their kids came to Maker Faire, it "blew their minds." It's not something they can see on TV or on their iPads. It's usually not the kind of experience they have at school. Mind-blowing experiences at a young age help children realize that life can be out-of-bounds, experimental, and exuberant. And we can expect them to want to participate once they have seen it for themselves.

Both Maker Faire and Burning Man are unfinished, edgy, and rough. They are strange rather than familiar, more original than a reproduction. You really don't know what to expect. Yet it becomes your own unique experience, not something you were told to experience. Such experiences can inspire us to tell our own story, sing our own song, build our own scrap-metal robot, or make our own art car. Both represent divergent thinking: there are just as many paths as there are people and their interests. And those paths can be connected by participants.

Ken Robinson talks about divergent thinking in his popular RSA Animate talk:

> Divergent thinking isn't the same thing as creativity. I define creativity as the process of having original ideas that have value. Divergent thinking isn't a synonym but is an essential capacity for creativity. It's the ability to see lots of possible answers to a question, lots of possible ways to interpret a question, to think laterally, to think not just in linear or convergent ways, to see multiple answers, not one.[10]

Consumer culture is based on what most people agree on, what most people will prefer, what most people will choose. It leads to consensus and conformity, which are also the hallmarks of public education and corporate marketing. In institutions, it is expressed as getting on the same page and rallying the troops. In schools, it means endless pages of textbooks and answering mind-numbing questions on a standardized test.

A future based on convergent thinking would forge consensus on what it is we will create. A future based on divergent thinking can explore many possible futures. We don't all need to be in agreement on what to do. Or perhaps it might be better to say that we seek to have only a minimal set of basic agreements, and yet provide for as many people as possible to participate and create.

Once, as I was getting on an elevator for a meeting at the Department of Education in Washington, D.C., a man was getting off and recognized me, although I didn't know him. He said he'd just been in a meeting where making came up as a topic. He asked me: "Is there a making project that every seventh grader in America could do?" Realizing that he wanted each student to do the same project, I shook my head and said, "No." I should have replied: "Yes, they should each do their own project." As MaKey MaKey inventor Jay Silver said, "We don't want everybody the same. We want a diverse ecosystem of unique thinkers. The Maker Movement isn't about robots or 3-D printers at all; it's really about freedom, the freedom for us with our hands to make the world we live in."

It is this sense of freedom that the Maker Movement holds as a moral obligation. Not the childish "It's a free country; I can say what I want and you can't stop me" kind of freedom espoused by libertarians. That's what John Dewey called the conventional notion of freedom, signifying "freedom from subjection to the will and control of others," which is just a precondition for what he calls the second sense of freedom, "effective freedom," not something which is given, but rather something that is *made*.[11] It is our exercising our ability to make and learn that is our means to realize true freedom. It's not just about achieving personal goals but working together with others in collaboration. The moral imperative of the Maker Movement is to use our creative freedom to make the future better, to be hands-on in making change, and to get everyone participating fully in that future.

EVERY CHILD A MAKER

The future will be made by our children, and so they deserve to be a primary focus of the Maker Movement. I have a young grandson, and watching him grow up is such a joy. Because of what I have learned about making and makers, I see young Henry differently than I saw my own children. I see him as a maker. I cherish every little thing he does, from stacking blocks to piling boxes so he can stand up to an adult-height counter. These are the signs that he's a maker. I see him figure things out for himself, and that he's always learning, always in the moment. I also realize how we might cultivate what is already present in him, support his development, and help him understand what he's capable of doing for himself and others. If the Maker Movement causes more parents and grandparents to recognize the pattern of a maker in their children, we will give those children a boost for the future.

What kind of life do we want for our children? I know what kind of life I wish for Henry: joyful, playful, experimental, expressive, resourceful, daring, uplifting, challenging, creative, and more. Recognizing the maker in every child and guiding the development of young makers is what we must do as adults to care for our children and our community's children.

Already I wonder, what will school be like for Henry and today's youngest children? Will his experience at school be different than my children's? Will school support his growth and development, or will school hinder it, as many schools do now? Will he be bored or passionate, made to sit still, or inspired to initiate creative acts?

I also wonder, will the world of work be different for these children? Will they be able to take advantage of their talents and continue developing them over the course of their lives? Will they feel that they are in control of their lives? Will they understand that the world can be hacked, tinkered with, taken apart, and put back together? That is what it means to get hands-on in making the world better.

Here's what I would want Henry and children everywhere to know:

Imagine the world being different because of you. Each of us can express who we are by what we learn to do and the things we make. Discover your creative talents and believe in using them. Learn technical skills and develop them through real-world practice. Find joy in what you are able to do for yourself and others. Join together to share learning, playing, working, and caring in our community and our culture. All of us are makers.

Appendix: A Chronology of the Maker Movement

2000 Douglas Repetto organizes first Dorkbot meeting

2001 Casey Reas and Benjamin Fry develop Processing, an open-source programming language

2002 MIT's Neil Gershenfeld starts Fab Lab

2003 Etsy launches marketplace for handmade goods
Nathan Seidle starts SparkFun Electronics
Bunnie Huang writes Hacking the Xbox
Leah Kramer launches craftster

2004 Kinetic Dress designed by CuteCircuit of London
Hernando Barragán develops Wiring project at IVREA in Italy

2005 First *Make:* magazine from O'Reilly Media uses term "maker"
Five-person Arduino team releases first board
Instructables website launches as DIY project instructions
Adafruit launched in Limor Fried's dorm room
Phil Torrone launches the Make: blog at makezine.com

2006 First Maker Faire in Bay Area
First TechShop opens in Menlo Park
First *Craft:* magazine published by O'Reilly Media
Leah Buechley designs Lillypad Arduino

2007 First RepRap Darwin 3D printer released
Chris Anderson starts DIY Drones website
Georgia Tech opens Invention Studio, a student-run makerspace
Karen Wilkinson and Mike Petrich open Tinkering Studio at Exploratorium
Fritzing for electronics design developed in Germany

2008 Eric Pan starts SEEED Studio in Shenzhen, China
DIYBio organization started
Noisebridge opens a hackerspace in SF
NYC Resistor opens in Brooklyn
Thingiverse 3D design repository launched

2009 Kickstarter launches crowdfunding platform for creative projects
Bre Pettis, Zach Smith and Adam Mayer form Makerbot in Brooklyn
Makerbot Cupcake CNC featured in *Make:* magazine
First Maker Faire in UK in Newcastle-upon-Tyne
Pumping Station One opens in Chicago as makerspace
First Mini Maker Faires in Rhode Island and Ann Arbor, Michigan
Maker Faire Africa in Accra, Ghana
Cory Doctorow publishes novel titled *Makers*

2010 Makerbot Thing-o-Matic introduced
Adafruit offers reward for hacking Kinect
Maker Faire Detroit opens
World Maker Faire in NYC opens
First Open Hardware Summit in New York City
Artisan's Asylum founded in Somerville, MA
David Li opens first hackerspace in China, Xinchejian
Genspace opens as DIYbio lab in Brooklyn
I3 Detroit opens as makerspace
Open Hardware Workshop at Eyebeam in New York City
First Open Source Hardware Definition published
Biocurious opens in Sunnyvale, CA as hackerspace for biotech
Limor Fried debuts weekly "Ask An Engineer" program

2011 Brook Drumm's Printrbot Kickstarter funded
Ayah Bdeir forms a company called LittleBits
Maker Works opens in Ann Arbor, MI
Dougherty gives TED talk "We Are Makers" in Detroit
Safecast project responds to Fukijima disaster
CERN publishes Open Hardware License (OHL)
Ultimaker founded in Netherlands to produce 3D printer
Jasen Wang starts MakeBlock in Shenzhen
Autodesk acquires Instructables
Tinkercad provides a browser-based 3D design platform
First FabLearn Conference Organized at Stanford by Paulo Blikstein

2012 First Raspberry Pi released
Makerbot Replicator introduced
Chris Anderson publishes *Makers*
FabCafe opens in Tokyo
Fayetteville, NY Free Public Library builds first library makerspace
Maker Ed as an educational nonprofit focused on making for kids
Groningen (Netherlands) is first European Maker Faire
Hong Kong Mini Maker is first Asian Faire
Joey Hudy brings marshmallow cannon to White House
First Maker Faire Tokyo
Make's first 3D Printer Shootout and Buyer's Guide
OpenBeam develops open-source construction kit
Autodesk develops free design apps as 123D

2013 Stratasys acquires Makerbot
First Maker Faire Rome
Intel releases Galileo board
First Maker Faire Shenzhen
Makerspace Playbook for high schools published
Autodesk acquires Tinkercad
Shopbot Tools introduces the Handibot

2014 White House Maker Faire celebrates American makers
Intel releases Edison board
GE FirstBuild opens in Louisville, Kentucky
NASA tests 3D printing on International Space Station
Local Motors develops 3D printed car

2015 Maker Faire Bay Area celebrates 10 years
Inventables launches Carvey, a 3D carving tool
Glowforge laser cutter launched on Kickstarter
National Week of Making organized by White House
New think[box] makerspace opens at Case Western Reserve University
MakerHealth Space opens in hospital in Galveston, TX

2016 European Maker Week organized by European Union
White House organizes second National Week of Making
Intel sponsors TV show *America's Greatest Makers*

Notes

INTRODUCTION

1. Interview with Gary Wills, PBS, https://www.pbs.org/jefferson
/archives/interviews/Wills.htm

1. WE ARE ALL MAKERS

1. Howard Gardner and Thomas Hatch, "Multiple Intelligences Go to
School: Educational Implications of the Theory of Multiple Intelligences," *Educational Researcher* 18:8 (November 1989), 5.
2. Yuval Noah Harari, *Sapiens: A Brief History of Humankind* (New
York: Harper, 2015).
3. Johan Huizinga, *Homo Ludens: A Study of the Play-Element in Culture*
(London: Routledge & Kegan Paul, 1949; Repr. London: Routledge
1998), ix.
4. Ibid., 3.
5. Ibid., 8.
6. Ibid., ix.
7. The film appeared in Detroit drive-in theaters during the run of Alfred
Hitchcock's *Psycho*. My friend Rick Prelinger of Prelinger Archives
(https://archive.org/details/prelinger) gave me a copy of this film.
8. Tom Wolfe, "The Tinkerings of Robert Noyce: How the Sun Rose
on the Silicon Valley," *Esquire* (December 1983), 346–74.
9. Dale Dougherty, "The Soul of an Old Heathkit," *Make:* 4 (2005),
72–73.
10. Flux is a cleaning agent used to prevent oxidation in metal soldering.
Because it burns before the solder, it's what you smell when soldering.
11. Benjamin Barber, *Consumed: How Markets Corrupt Children, Infantilize Adults, and Swallow Citizens Whole* (New York: W. W. Norton,
2007), 139.

12. Harari, *Sapiens*, 43.
13. Quoted by Sarah Nicole Prickett, "Patti Smith: The Responsible Artist," *The Globe and Mail* (Toronto), June 11, 2012.
14. Steven Levy, *Hackers: Heroes of the Computer Revolution* (New York: Anchor, 1984), 3.
15. Stephen Wozniak, "Homebrew and How the Apple Came to Be," in *Digital Deli: The Comprehensive, User-Lovable Menu of Computer Lore, Culture, Lifestyles, and Fancy*, ed. Steve Ditlea (New York: Workman, 1984), www.atariarchives.org/deli/homebrew_and_how_the_apple .php.
16. Ibid.
17. Eric S. Raymond, *The Cathedral and the Bazaar: Musings on Linux and Open Source by an Accidental Revolutionary* (Sebastopol, CA: O'Reilly Media, 1999), 23.
18. David H. Ahl, "The First West Coast Computer Faire," in *The Best of Creative Computing*, vol. 3 (Morristown, NJ: Creative Computing Press, 1980), www.atariarchives.org/bcc3/showpage .php?page=98.
19. Ibid.
20. The name is a bit of an inside joke. I got the name *Make* from a UNIX utility program, which was the topic of an early O'Reilly book, *Managing Projects with Make*. This utility compiles a list of programs, using a file that starts off with "Make:."

2. WHO: AMATEURS, ENTHUSIASTS, AND PROFESSIONALS

1. Charles Leadbeater and Paul Miller, "The Pro-Am Revolution: How Enthusiasts Are Changing Our Economy and Society," white paper (London: Demos, 2004), 20.
2. Ibid.
3. Richard Holmes, *The Age of Wonder: How the Romantic Generation Discovered the Beauty and Terror of Science* (New York: Pantheon, 2009), 87.
4. Ibid., 95.
5. Ibid., 79.

6. Forrest M. Mims III, "Country Scientist: Becoming an Amateur Scientist," *Make:* 24 (October 2010), 26–28.
7. John Brockman, "The Technium: a Conversation with Kevin Kelly," Edge.org, February 3, 2014, http://edge.org/conversation/the-technium.
8. David Gauntlett, *Making Is Connecting: The Social Meaning of Creativity, from DIY and Knitting to YouTube and Web 2.0* (Cambridge, UK: Polity Press, 2011), 12.
9. Carla Sinclair, "Welcome," *Craft:* 1 (October 2006), 7.
10. Ulla-Maaria Mutanen, "Crafter Manifesto," *Make:* 4 (2005), 7.
11. National Fairground Archive, University of Sheffield, www.sheffield.ac.uk/nfa.

3. WHAT: ART, INTERACTION, AND INNOVATION

1. Make: Television, "Kite Aerial Photography," January 2009, www.youtube.com/watch?v=swqFA9Mvq5M.
2. U.S. Geological Survey, Earthquake Hazards Program, "Rephotographing George Lawrence's 'San Francisco in Ruins,'" July 2012, http://earthquake.usgs.gov/regional/nca/1906/kap.
3. Cris Benton, *Saltscapes: The Kite Aerial Photography of Cris Benton* (Berkeley, CA: Heyday, 2013).
4. Ibid., 2.
5. Eric Wilhelm, "How to Start a Business," Instructables.com, www.instructables.com/id/How-to-Start-a-Business-1.
6. Amy Cuddy, *Presence: Bringing Your Boldest Self to Your Biggest Challenges* (New York: Hachette, 2015); Amy Cuddy, "Your Body Language Shapes Who You Are," TED Talk, October 2012, www.ted.com/talks/amy_cuddy_your_body_language_shapes_who_you_are.
7. Scott Heimendinger, "Sous-Vide Immersion Cooker," *Make:* 25 (February 2011), 107.
8. Ibid.
9. Eric von Hippel, *Democratizing Innovation* (Cambridge, MA: MIT Press, 2005), 93.

4. WHERE: COMMUNITIES, SCHOOLS, AND INDUSTRY

1. Jeremy Rifkin, *The Zero Marginal Cost Society* (New York: St. Martin's, 2014), 21.
2. Georgia Guthrie, "Where Are the Women?," *Make:* 40 (July 2014), 53.
3. Neil Gershenfeld, "Welcome to the Fab Lab," *Make:* 1 (February 2005), 25.
4. Nugget ice, often used in sodas in restaurants, takes the form of tiny, soft, chewable pellets. This is an affordable nugget ice maker for the home.
5. "Maker Culture to Be Encouraged in China," *Want China Times,* March 15, 2015.
6. "Shenzhen to Be Built as City of 'Makers,'" *China Daily,* April 19, 2015, www.china.org.cn/china/2015-04/19/content_35360655.htm.
7. Maker Nation, "2014 State of Makespaces," www.slideshare .net/TheMakersNation/fab10-state-of-makerspaces-survey-results -the-makers-nation.

5. HOW: COMPONENTS, TOOLS, AND MARKETS

1. Marshall McLuhan, *Understanding Media: The Extensions of Man* (1964; repr. Cambridge, MA: MIT Press, 1994), xxi.
2. Andrew "Bunnie" Huang, "Akihabara, Eat Your Heart Out," Bunnie:studios blog, January 31, 2007, www.bunniestudios.com /blog/?p=147.
3. Open Source Hardware Association, "Open Source Hardware (OSHW) Statement of Principles 1.0," www.oshwa.org /definition.
4. MakerBot was acquired by Stratasys Limited, "the leader in 3-D printing and additive manufacturing." Kelly Clay, "3-D Printing Company MakerBot Acquired in $604 Million Deal," *Forbes,* June 19, 2013, www.forbes.com/sites/kellyclay/2013/06/19/3d-printing -company-makerbot-acquired-in-604-million-deal/#2715e4857a0b 6f0cba45ff63.
5. Fred Wilson, "You are not your work," http://avc.com/2014/01 /you-are-not-your-work/, January 21, 2014.

6. Pam Klainer, "Rob Wilson and Etsy.com," https://pklainer.com/2009/10/22/rob-kalin-and-etsy-com/, Oct 22, 2009.

7. Hannah Miller, "How to Earn $1000s as a Micro-Entrepreneur Starting Now," Shareable, December 13, 2012, www.shareable.net/blog/how-to-earn-1000s-as-a-micro-entrepreneur-starting-now.

6. TOY MAKERS

1. Christiana Yambo and Sabastian Boaz, "Circuit-Bend Your Casio SK Keyboard," *Make:* 4 (2005), 88–101.

2. Ibid., 92.

3. Ayah Bdeir and Matthew Richardson, *Getting Started with LittleBits: Prototyping and Inventing with Modular Electronics* (Sebastopol, CA: Maker Media, 2015), ix.

4. Ibid., vii.

5. Mitchel Resnick, "Rethinking Learning in the Digital Age," in *The Global Information Technology Report 2001–2002: Readiness for the Networked World,* Geoffrey Kirkman and Klaus Schwab, eds. (New York: Oxford University Press, 2004), 33.

6. Elizabeth Sweet, "Beyond the Blue and Pink Toy Divide," TEDx, June 30, 2015, http://tedxtalks.ted.com/video/Beyond-the-Blue-and-Pink-Toy-Di.

7. Anne Mayoral, "What Is a 'Girl Toy'?," *Make:* 41 (September 2014).

7. THE MAKER

1. Mihaly Csikszentmihalyi, *Flow: The Psychology of Optimal Experience* (New York: Harper & Row, 1990), 6.

2. Carol S. Dweck, *Mindset: The New Psychology of Success* (New York: Random House, 2006), 6.

3. Ibid., 7.

4. Kelly Lambert, *Lifting Depression: A Neuroscientist's Hands-On Approach to Activating Your Brain's Healing Power* (New York: Basic Books, 2008), 6–7.

5. Ibid., 7.

6. Ibid., 33.

7. Stuart Brown, *Play: How It Shapes the Brain, Opens the Imagination, and Invigorates the Soul* (New York: Avery, 2009), 5.

8. Ibid., 11.

9. Ibid., 127–28.

10. Ibid., 126.

11. Shunryu Suzuki and David Chadwick, *Zen Mind, Beginner's Mind* (Boston: Shambhala, 2006).

12. Charles Eames and Ray Eames, *An Eames Anthology: Articles, Film Scripts, Interviews, Letters, Notes, and Speeches,* ed. Daniel Ostroff (New Haven, CT: Yale University Press, 2015).

13. William Lidwell, "The Dean of Engineering," *Make:* 4 (2005), 28–37.

14. Steven L. Goldman, "Why We Need a Philosophy of Engineering: A Work in Progress," *Interdisciplinary Science Reviews* 29:2 (2004), 163–76.

15. Craig R. Forest, Roxanne A. Moore, Amit S. Jariwala, Barbara Burks Fasse, Julie Linsey, Wendy Newstetter, Peter Ngo, and Christopher Quintero, "The Invention Studio: A University Maker Space and Culture," *Advances in Engineering Education* 4:2 (fall 2014), http://advances.asee.org/wp-content/uploads/vol04/issue02/papers/AEE-14-1-Forest.pdf.

16. Goldman, "Philosophy of Engineering."

17. Todd Lappin, "A Feel for Engineering," *Make:* 19 (July 2009), 30–35.

18. Karl Popper, *All Life Is Problem Solving* (Abingdon, UK: Routledge, 1999), 100.

19. Stett Holbrook, "Audacious by Design: Project H Gives Kids Tools, Skills, and Confidence," *Make:* 40 (July 2014), 20–23.

20. Jay Silver, "The Maker Movement Is Not about Robots," YouTube, October 27, 2015, www.youtube.com/watch?v=Sa9vgqz7hMs.

21. Frank Bidart, "Advice to the Players," *Star Dust* (New York: Farrar, Straus and Giroux, 2006), 10.

22. Walter Isaacson, "The America Ben Franklin Saw," *Washington Post*, November 11, 2012.

23. Eliot Wigginton and Paul F. Gillespie, *The Foxfire Book* (Garden City, NY: Anchor Press/Doubleday, 1972), 20.

24. Ibid., 30.

25. Ibid., 38.

26. Eugene S. Ferguson, *Oliver Evans, Inventive Genius of the American Industrial Revolution* (Wilmington, DE: Hagley Museum, 1980), 30.

27. *The Young Mill-wright and Miller's Guide,* https://goo.gl/5riXmv.

28. Boston Women's Health Book Collective, *Our Bodies, Ourselves: A Book by and for Women* (New York: Simon and Schuster, 1973).

29. Stewart Brand, *Whole Earth Catalog* 1 (fall 1968).

30. Kevin Kelly, talk at MakerCon Bay Area, May 2014, YouTube, www.youtube.com/watch?v=IhF6wDmwAmk.

31. Marvin Minsky, *The Society of Mind* (New York: Simon & Schuster, 1987), 2.

8. MAKING IS LEARNING

1. Karen Wilkinson and Mike Petrich, *The Art of Tinkering* (San Francisco: Weldon Owen, 2014), 13.

2. Brown, *Play,* 105.

3. Tim Walker, "The Joyful, Illiterate Kindergartners of Finland," *The Atlantic,* October 1, 2015.

4. David Elkind, "Can We Play?," University of California, Berkeley, Greater Good, March 1, 2008, http://greatergood.berkeley.edu /article/item/can_we_play.

5. Curt Gabrielson, *Tinkering: Kids Learn by Making Stuff* (Sebastopol, CA: Maker Media, 2015), ix.

6. Ibid.

7. James P. Spillane, "Data in Practice: Conceptualizing the Data-Based Decision-Making Phenomena," *American Journal of Education* 118:2 (February 2012), 113–41.

8. Lisa Brahms and Peter Wardrip, "The Learning Practices of Making: An Evolving Framework for Design," Children's Museum of Pittsburgh, December 2014.

9. Maria Montessori, *The Absorbent Mind* (New York: Henry Holt, 1967), 167.

10. John Dewey, *Democracy and Education: An Introduction to the Philosophy of Education* (New York: Macmillan, 1916; repr. New York: Free Press, 1966), 112.

11. John Dewey, "My Pedagogic Creed," *The School Journal* 54 (January 1897), 78.

12. Jean Piaget, *To Understand Is to Invent* (New York: Grossman, 1973), 20.

13. Jerome S. Bruner, *Toward a Theory of Instruction* (Cambridge, MA: Harvard University Press, 1966), 69.

14. Robert Halpern, Paul Heckman, and Reed Larson, "Realizing the Potential of Learning in Middle Adolescence," January 2013, www.realizinglearning.org.

15. Piaget, *To Invent*, 44.

16. Sylvia Libow Martinez and Gary Stager, *Invent to Learn: Making, Tinkering, and Engineering in the Classroom* (Torrance, CA: Constructing Modern Knowledge Press, 2013), 3.

17. Agency by Design at Project Zero, "Maker-Centered Learning and the Development of Self," Harvard University Graduate School of Education, January 2015, 3–4, www.pz.harvard.edu/sites/default/files/Maker-Centered-Learning-and-the-Development-of-Self_AbD_Jan-2015.pdf.

18. Ibid., 4.

9. MAKING IS WORKING

1. Jane Jacobs, *The Economy of Cities* (New York: Vintage, 1970), 121.

2. National Association of Manufacturers, "Top 20 Facts about Manufacturing," www.nam.org/Newsroom/Top-20-Facts-About-Manufacturing.

3. Danit Peleg, http://danitpeleg.com.

4. Records of the Falls City Brewing Company, housed in the University of Louisville Archives, http://louisville.edu/library/archives.

5. Peter F. Drucker, *The Practice of Management* (New York: Harper & Row, 1954), 34.
6. Erik Kain, "The Rise of Craft Beer in America," *Forbes,* September 16, 2011, www.forbes.com/sites/erikkain/2011/09/16/the-rise-of-craft-beer-in-america/#6ffdd9d27d79.
7. Brewers Association, "Craft Brewing Statistics," www.craftbeer.com/breweries/support-your-local-brewery/craft-brewing-statistics.
8. Opendesk, www.opendesk.cc.
9. Eames and Eames, *Eames Anthology;* see also http://eamesfoundation.org.
10. Micah Lande, "Catalysts for Design Thinking and Engineering Thinking: Fostering Ambidextrous Mindsets for Innovation," Mudd Design Workshop IX, Harvey Mudd College, Claremont, CA, May 2015.
11. Hal R. Varian, "The 2009 Guglielmo Marconi Lecture," The Lisbon Council, June 2009, www.youtube.com/watch?v=hqaA-fgdXEE.
12. Carl Benedikt Frey and Michael A. Osborne, "The Future of Employment: How Susceptible Are Jobs to Computerization?" University of Oxford, September 17, 2013, www.oxfordmartin.ox.ac.uk/downloads/academic/The_Future_of_Employment.pdf.
13. "Protest against Use of Typesetting Machines," *Los Angeles Herald,* December 28, 1903.

10. MAKING IS CARING

1. ErgoJoystick, www.ergojoystick.com.
2. Nicolas Huchet, "How I Made My $250 Robotic Arm," *Make:* 43 (January 2015), http://makezine.com/2015/01/22/how-made-my-250-robotic-arm.
3. José Gómez-Márquez, "Design for Hack in Medicine: MacGyver Nurses and Legos Are Helping Us Make MEDIKits for Better Health Care," *Make: Ultimate Kit Guide,* 2012.
4. MakerHealth, www.makerhealth.co.
5. Gómez-Márquez, "Design for Hack."

6. José Gómez-Márquez, "A Hospital Mini Maker Faire," *Make:* blog, May 30, 2014.

7. Dana Lewis, "Building Diabetes Technology Is like Building a Mountain Bike," DIPYS.org blog, October 29, 2015, http://diyps .org.

8. Ariane Conrad, "We Are All Crew," Medium, September 11, 2015, https://medium.com/poc-stories/we-are-all-crew -68b10a4b0819#.8h4nxlwg5.

9. Jane Addams, *Twenty Years at Hull-House* (New York: Macmillan, 1911), 75.

10. Ibid., 196.

11. Ibid., 181.

12. Ibid., 74.

13. DJ Jazzy Jay, NAMM interview, November 12, 2012, www.namm. org/library/oral-history/dj-jazzy-jay.

14. Maker Education Initiative, http://makered.org.

15. Irwin Kula, "Religion as Tech: Applying Innovation Theory to Making 'Good People,'" Business Innovation Factory, September 2014, www.businessinnovationfactory.com/summit/video/irwin -kula-religion-tech-applying-innovation-theory-making-good -people#.VvwVPPkrJGo.

11. MAKING THE FUTURE

1. Isaac Asimov, "Visit to the World's Fair of 2014," *New York Times,* August 16, 1964.

2. Ibid.

3. White House, "Computer Science for All," January 30, 2016, www .whitehouse.gov/blog/2016/01/30/computer-science-all.

4. Asimov, "Visit to the World's Fair."

5. Patrick O'Connor, "Poll Finds Widespread Economic Anxiety," *Wall Street Journal,* August 5, 2014.

6. Mortimer Zuckerman, "The End of American Optimism," *Wall Street Journal,* August 16, 2010.

7. Pew Research Center, "Emerging and Developing Economies

Much More Optimistic than Rich Countries about the Future," October 9, 2014, www.pewglobal.org/2014/10/09/emerging -and-developing-economies-much-more-optimistic-than -rich-countries-about-the-future.

8. Henry Jenkins, "Confronting the Challenges of Participatory Culture: Media Education for the 21st Century," white paper for the MacArthur Foundation (Cambridge, MA: MIT Press, 2009), 7.

9. *Ear Candy,* December 2002, www.earcandymag.com.

10. Ken Robinson, "Changing Education Paradigms," RSA Animate, October 2010, www.youtube.com/watch?v=zDZFcDGpL4U.

11. John Dewey and James Tufts, "Responsibility and Freedom," in *Ethics* (New York: Henry Holt, 1909), 437.

Bibliography

Addams, Jane. *Twenty Years at Hull-House.* New York: Macmillan, 1911.

Agency by Design at Project Zero. "Maker-Centered Learning and the Development of Self." Harvard University Graduate School of Education. January 2015. www.pz.harvard.edu/sites/default/files/ Maker-Centered-Learning-and-the-Development-of-Self_AbD_Jan-2015.pdf.

Ahl, David H. "The First West Coast Computer Faire." In *The Best of Creative Computing,* vol. 3. Morristown, NJ: Creative Computing Press, 1980, www.atariarchives.org/bcc3/showpage.php?page=98.

Anderson, Chris. *Makers: The New Industrial Revolution.* New York: Crown Business, 2012.

Appius Claudius Caecus. *Sententiae.* Circa 300 BCE.

Asimov, Isaac. "Visit to the World's Fair of 2014." *New York Times,* August 16, 1964.

Barber, Benjamin. *Consumed: How Markets Corrupt Children, Infantilize Adults, and Swallow Citizens Whole.* New York: W. W. Norton, 2007.

Bdeir, Ayah, and Matthew Richardson. *Getting Started with LittleBits: Prototyping and Inventing with Modular Electronics.* Sebastopol, CA: Maker Media, 2015.

Benton, Cris. *Saltscapes: The Kite Aerial Photography of Cris Benton.* Berkeley, CA: Heyday, 2013.

Bidart, Frank. *Star Dust.* New York: Farrar, Straus and Giroux, 2006.

Boston Women's Health Book Collective. *Our Bodies, Ourselves: A Book by and for Women.* New York: Simon and Schuster, 1973.

Brahms, Lisa, and Peter Wardrip. "The Learning Practices of Making: An Evolving Framework for Design." Children's Museum of Pittsburgh. December 2014.

Bringuier, Jean-Claude. *Conversations with Jean Piaget.* Translated by Basla Mille Gulati. Chicago: University of Chicago Press, 1980.

Brockman, John. "The Technium: a Conversation with Kevin

Kelly." Edge.org. February 3, 2014. http://edge.org/conversation /the-technium.

Brown, Stuart. *Play: How It Shapes the Brain, Opens the Imagination, and Invigorates the Soul.* New York: Avery, 2009.

Bruner, Jerome S. *Toward a Theory of Instruction.* Cambridge, MA: Harvard University Press, 1966.

Campanella, Roy. *It's Good to Be Alive.* Boston: Little Brown & Co., 1959.

Creadon, Patrick. *If You Build It.* Documentary film. 2013. www .ifyoubuilditmovie.com.

Csikszentmihalyi, Mihaly. *Flow: The Psychology of Optimal Experience.* New York: Harper & Row, 1990.

Cuddy, Amy. "Your Body Language Shapes Who You Are." TED Talk. October 2012. www.ted.com/talks/amy_cuddy _your_body_language_shapes_who_you_are.

Dewey, John. *Democracy and Education: An Introduction to the Philosophy of Education.* New York: Macmillan, 1916. Reprinted New York: Free Press, 1966.

———. "My Pedagogic Creed." *The School Journal* 54 (January 1897).

Dewey, John, and James Tufts. *Ethics.* New York: Henry Holt, 1909.

Drexler, K. Eric. *Engines of Creation: The Coming Era of Nanotechnology.* New York: Bantam Doubleday Dell, 1986.

Drucker, Peter F. *The Practice of Management.* New York: Harper & Row, 1954.

Dweck, Carol S. *Mindset: The New Psychology of Success.* New York: Random House, 2006.

Eames, Charles, and Ray Eames. *An Eames Anthology: Articles, Film Scripts, Interviews, Letters, Notes, and Speeches.* Edited by Daniel Ostroff. New Haven, CT: Yale University Press, 2015.

Eliade, Mircea. *The Forge and the Crucible.* Translated by Stephen Corrin. New York: Harper and Row, 1971. Originally published as *Forgerons et alchimistes* (Paris: Flammarion, 1956).

Ferguson, Eugene S. *Oliver Evans, Inventive Genius of the American Industrial Revolution.* Wilmington, DE: Hagley Museum, 1980.

Forest, Craig R., Roxanne A. Moore, Amit S. Jariwala, Barbara Burks

Fasse, Julie Linsey, Wendy Newstetter, Peter Ngo, and Christopher Quintero. "The Invention Studio: A University Maker Space and Culture." *Advances in Engineering Education* 4:2 (fall 2014). http://advances.asee.org/wp-content/uploads/vol04/issue02/papers/AEE-14-1-Forest.pdf.

Franklin, Benjamin. *The Autobiography of Benjamin Franklin.* London: J. Parson's, 1793.

Frey, Carl Benedikt, and Michael A. Osborne. "The Future of Employment: How Susceptible Are Jobs to Computerization?" University of Oxford. September 17, 2013. www.oxfordmartin.ox.ac.uk/downloads/academic/The_Future_of_Employment.pdf.

Gabrielson, Curt. *Tinkering: Kids Learn by Making Stuff.* Sebastopol, CA: Maker Media, 2015.

Gardner, Howard, and Thomas Hatch. "Multiple Intelligences Go to School: Educational Implications of the Theory of Multiple Intelligences." *Educational Researcher* 18:8 (November 1989), 4–10.

Gauntlett, David. *Making Is Connecting: The Social Meaning of Creativity, from DIY and Knitting to YouTube and Web 2.0.* Cambridge, UK: Polity Press, 2011.

Gershenfeld, Neil. *Fab: The Coming Revolution on Your Desktop—from Personal Computers to Personal Fabrication.* New York: Basic Books, 2008.

Goldman, Steven L. "Why We Need a Philosophy of Engineering." *Interdisciplinary Science Reviews* 29:2 (2004), 163–76.

Gurstelle, William. *Backyard Ballistics: Build Potato Cannons, Paper Match Rockets, Cincinnati Fire Kites, Tennis Ball Mortars, and More Dynamite Devices.* Chicago: Chicago Review Press, 2001.

Hagel, John, John Seely Brown, and Lang Davison. *The Power of Pull: How Small Moves, Smartly Made, Can Set Big Things in Motion.* New York: Basic Books, 2010.

Halpern, Robert, Paul Heckman, and Reed Larson. "Realizing the Potential of Learning in Middle Adolescence." January 2013. www.realizinglearning.org.

Harari, Yuval Noah. *Sapiens: A Brief History of Humankind.* New York: Harper, 2015.

Hatch, Mark. *The Maker Movement Manifesto.* New York: McGraw

Hill Education, 2013.

Holmes, Richard. *The Age of Wonder: How the Romantic Generation Discovered the Beauty and Terror of Science.* New York: Pantheon, 2009.

Huang, Andrew "Bunnie." *Hacking the Xbox: An Introduction to Reverse Engineering.* San Francisco: No Starch Press, 2003.

Huizinga, Johan. *Homo Ludens: A Study of the Play-Element in Culture.* London: Routledge & Kegan Paul, 1949. Reprinted London: Routledge, 1998.

Isaacson, Walter. "The America That Ben Franklin Saw." *Washington Post,* November 21, 2012.

Jacobs, Jane. *The Economy of Cities.* New York: Vintage, 1970.

Jenkins, Henry. "Confronting the Challenges of Participatory Culture: Media Education for the 21st Century." White paper for the MacArthur Foundation. Cambridge, MA: MIT Press, 2009.

Jenkins, Henry, and Mizuku Ito. *Participatory Culture in a Networked Era: A Conversation on Youth, Learning, Commerce and Politics.* Cambridge, UK: Polity Press, 2015.

Jobs, Steve. 2005 Commencement Address at Stanford University. http://news.stanford.edu/news/2005/june15/jobs-061505.html.

Kelly, Kevin. *What Technology Wants.* New York: Viking, 2010.

Kirkman, Geoffrey, and Klaus Schwab, editors. *The Global Information Technology Report 2001–2002: Readiness for the Networked World.* New York: Oxford University Press, 2004.

Lambert, Kelly. *Lifting Depression: A Neuroscientist's Hands-On Approach to Activating Your Brain's Healing Power.* New York: Basic Books, 2008.

Lang, David. *Zero to Maker: Learn (Just Enough) to Make (Just About) Anything.* Sebastopol, CA: Maker Media, 2013.

Leadbeater, Charles, and Paul Miller. "The Pro-Am Revolution: How Enthusiasts Are Changing Our Economy and Society." White paper. London: Demos, 2004.

Levy, Steven. *Hackers: Heroes of the Computer Revolution.* New York: Anchor, 1984.

Maker Nation. "2014 State of Makerspaces." www.slideshare.net/TheMakersNation/fab10-state-of-makerspaces-survey-results-the-makers-nation.

Martinez, Sylvia Libow, and Gary Stager. *Invent to Learn: Making, Tinkering, and Engineering in the Classroom.* Torrance, CA: Constructing Modern Knowledge Press, 2013.

McLuhan, Marshall. *Understanding Media: The Extensions of Man.* Cambridge, MA: MIT Press, 1994. First published 1964 by McGraw-Hill.

Melville, Herman. "Bartleby, the Scrivener: A Story of Wall Street." *Putnam's Magazine,* November 1853.

Mims, Forrest M., III. "Amateur Science: Strong Tradition, Bright Future," *Science* 284:5411 (April 2, 1999).

Minsky, Marvin. *The Society of Mind.* New York: Simon & Schuster, 1987.

Montessori, Maria. *The Absorbent Mind.* New York: Henry Holt, 1967.

Myhrvold, Nathan, Maxime Bilet, and Melissa Lehuta. *Modernist Cuisine at Home.* Bellevue, WA: Cooking Lab, 2012.

Papert, Seymour. *The Children's Machine: Rethinking School in the Age of the Computer.* New York: Basic Books, 1993.

Piaget, Jean. *To Understand Is to Invent.* New York: Grossman, 1973.

Popper, Karl. *All Life Is Problem Solving.* Abingdon, UK: Routledge, 1999.

Raymond, Eric S. *The Cathedral and the Bazaar: Musings on Linux and Open Source by an Accidental Revolutionary.* Sebastopol, CA: O'Reilly Media, 1999.

Rifkin, Jeremy. *The Third Industrial Revolution: How Lateral Power Is Transforming Energy, the Economy, and the World.* New York: St. Martin's, 2011.

———. *The Zero Marginal Cost Society: The Internet of Things, the Collaborative Commons, and the Eclipse of Capitalism.* New York: St. Martin's, 2014.

Robinson, Ken. "Changing Education Paradigms." RSA Animate. October 2010. www.youtube.com/watch?v=zDZFcDGpL4U.

Seymour, John. *The Self-Sufficient Gardener.* London: Dorling Kindersley, 1978.

Silver, Jay. "The Maker Movement Is Not about Robots." YouTube. October 27, 2015. www.youtube.com/watch?v=Sa9vgqz7hMs.

Sinclair, Carla. *CRAFT: Transforming Traditional Crafts.* Sebastopol, CA: O'Reilly Media, 2008.

Spillane, James P. "Data in Practice: Conceptualizing the Data-Based Decision-Making Phenomena." *American Journal of Education* 118:2 (February 2012), 113–41.

Stallman, Richard M. *Free Software Free Society: Selected Essays of Richard M. Stallman.* 2nd ed. Edited by Joshua Gay. Cambridge, MA: Free Software Foundation, 2010.

Suzuki, Shunryu, and David Chadwick. *Zen Mind, Beginner's Mind.* Boston: Shambhala, 2006.

Sweet, Elizabeth. "Beyond the Blue and Pink Toy Divide." TEDx. June 30, 2015. http://tedxtalks.ted.com/video/Beyond-the -Blue-and-Pink-Toy-Di.

Usher, Abbott Payson. *A History of Mechanical Invention.* Rev. ed. Cambridge, MA: Harvard University Press, 1954.

Varian, Hal R. "The 2009 Guglielmo Marconi Lecture." The Lisbon Council. June 2009. www.youtube.com/watch?v=hqaA-fgdXEE.

Von Hippel, Eric. *Democratizing Innovation.* Cambridge, MA: MIT Press, 2005.

Walker, Tim. "The Joyful, Illiterate Kindergartners of Finland." *The Atlantic,* October 1, 2015.

WALL-E. Film. Pixar, 2008.

Watson, Bruce. *The Man Who Changed How Boys and Toys Were Made.* New York: Viking Adult, 2002.

Wigginton, Eliot, and Paul F. Gillespie. *The Foxfire Book.* Garden City, NY: Anchor/Doubleday, 1972.

Wilkinson, Karen, and Mike Petrich. *The Art of Tinkering.* San Francisco: Weldon Owen, 2014.

Wolfe, Tom. "The Tinkerings of Robert Noyce: How the Sun Rose on the Silicon Valley." *Esquire,* December 1983, 346–74.

Wozniak, Stephen. "Homebrew and How the Apple Came to Be." In *Digital Deli: The Comprehensive, User-Lovable Menu of Computer Lore, Culture, Lifestyles, and Fancy.* Edited by Steve Ditlea. New York: Workman, 1984. www.atariarchives.org/deli/homebrew_and_how _the_apple.php.

Zuckerman, Mortimer. "The End of American Optimism." *Wall Street Journal,* August 16, 2010.

Index

G

About the Authors

Dale Dougherty is the founder and CEO of Maker Media Inc. in San Francisco. Maker Media produces *Make:* magazine, which launched in 2005, and the Maker Faire, which was held first in the San Francisco Bay Area in 2006. There were 151 Maker Faires held around the world in 2015. Dougherty was born in 1955 in Los Angeles and grew up in Louisville, Kentucky. He was a cofounder of O'Reilly Media, where he was the first editor of their computing trade books, and developed GNN, the first commercial website, in 1993. He coined "Web 2.0" in 2003. *Make:* started at O'Reilly Media and spun out as its own company in January 2013. In 2011 Dougherty was honored at the White House as a Champion of Change through an initiative that honors Americans who are "doing extraordinary things in their communities to out-innovate, out-educate, and out-build the rest of the world." At the 2014 White House Maker Faire he was introduced by President Obama as an American innovator making significant contributions to the fields of education and business. He lives in Sebastopol, California, with his wife, Nancy.

Since 2007, Ariane Conrad, a freelance writer, editor, and coach known as the Book Doula, has collaboratively authored seven nonfiction books, including three *New York Times* best-sellers. Most recently she supported Kennedy Odede and Jessica Posner with *Find Me Unafraid: Love, Loss and Hope in an African Slum* (Ecco, 2015). More about her collaborations, interviews, presentations, and other adventures is available at http://arianeconrad.com.